This Astronomy Observation Planner

BELONGS TO:

____

DEDICATION

This Astronomy Observation Planner is dedicated to all the ones who love the night sky and would like to document their findings in the process.

You are my inspiration for producing books and I'm honored to be a part of keeping all of your information, notes and records organized all in one easy to find spot.

How to use this Astronomy Observation Planner:

This ultimate Astronomy Observation notebook is a perfect way to track and record all your stargazing activities. This unique Astronomy logbook is a great way to keep all of your important information all in one place.

Each interior page includes prompts and space to record the following:

1. Date - Write the date, time, location, GPS and observer.

2. Sky Conditions - Use the description timeline to indicate from clear skies to misty.

3. Equipment/Tools - Stay on task using the space to fill in telescopes, spectrographs, cameras, computers used to observe objects in the universe.

4. Finder - Record and track with this hand drawn map of a small region of the sky you are observing.

5. EP/MAG/Filter - Describe how bright the objects are using the magnitude scale.

If you are new to the field of Stargazing or have been at it for a while, this Astronomy Observation Journal is a must have! Can make a great useful gift for anyone who loves observing the sky! Convenient size of 8x10 inches, 110 pages, quality white paper, soft matte finish cover, paperback.

Have Fun!

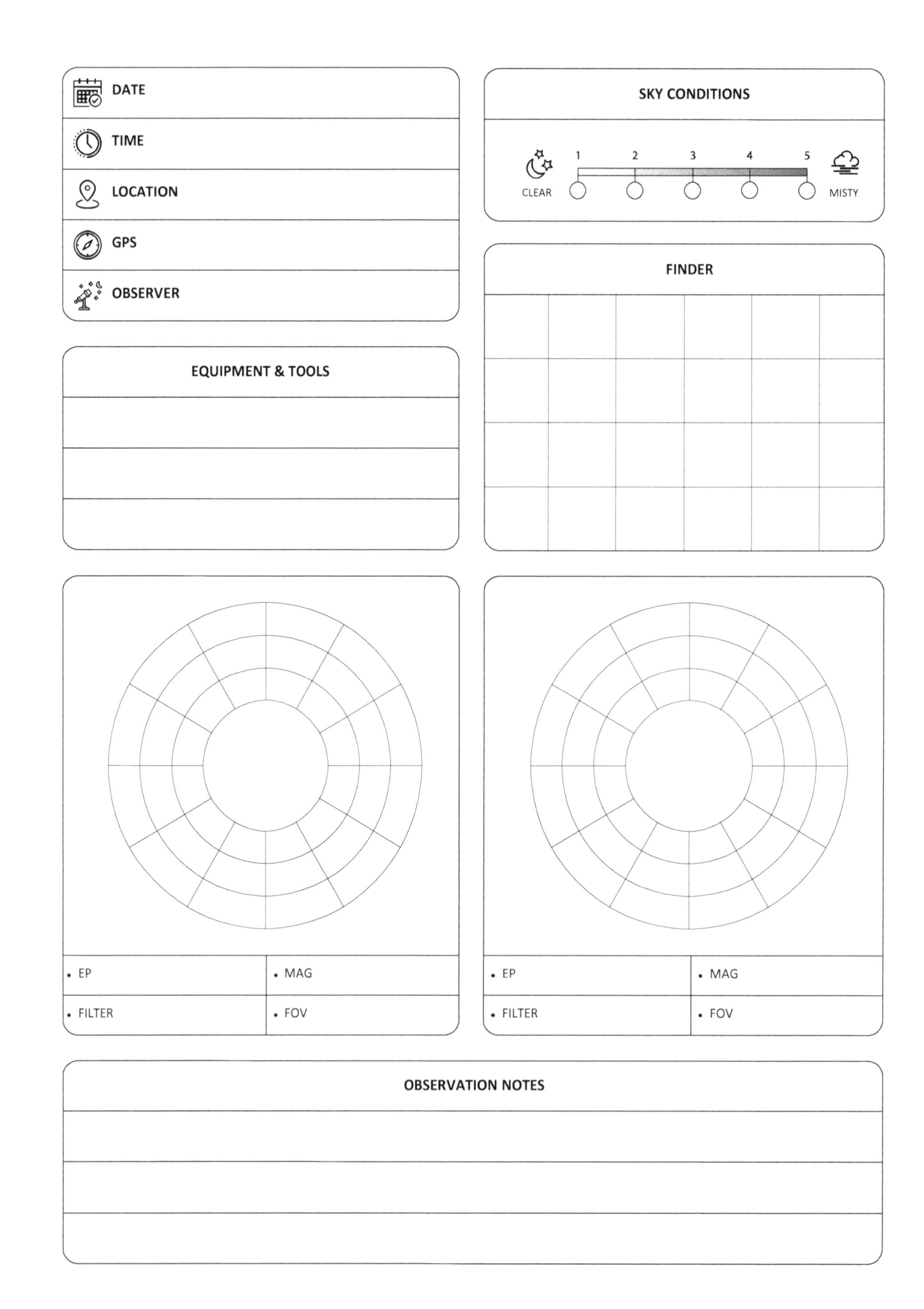
DATE
TIME
LOCATION
GPS
OBSERVER
SKY CONDITIONS
1
2
3
4
5
CLEAR
MISTY
FINDER
EQUIPMENT & TOOLS
EP
MAG
FILTER
FOV
EP
MAG
FILTER
FOV
OBSERVATION NOTES

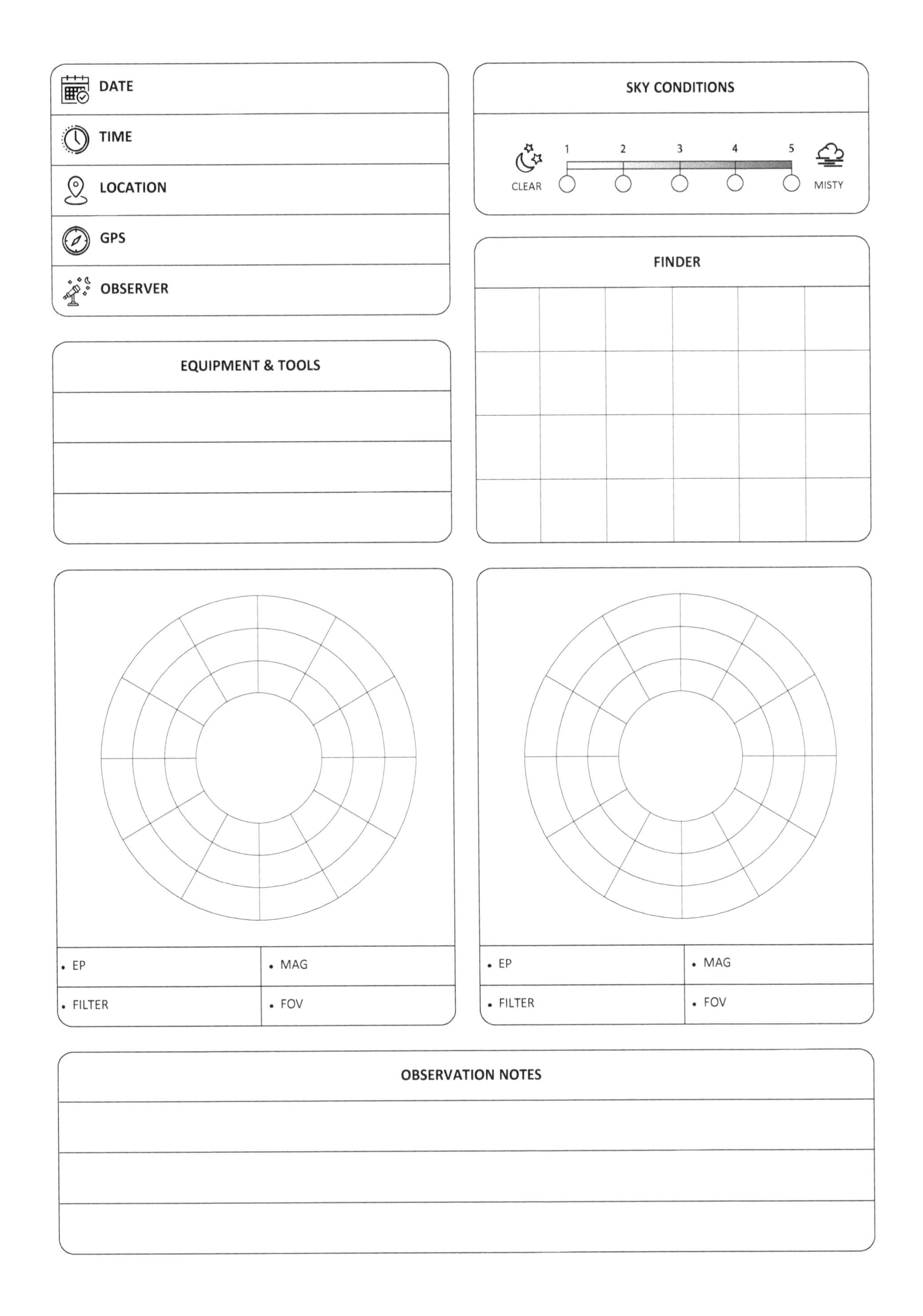
DATE
TIME
LOCATION
GPS
OBSERVER
SKY CONDITIONS
1
2
3
4
5
CLEAR
MISTY
FINDER
EQUIPMENT & TOOLS
EP
MAG
FILTER
FOV
EP
MAG
FILTER
FOV
OBSERVATION NOTES

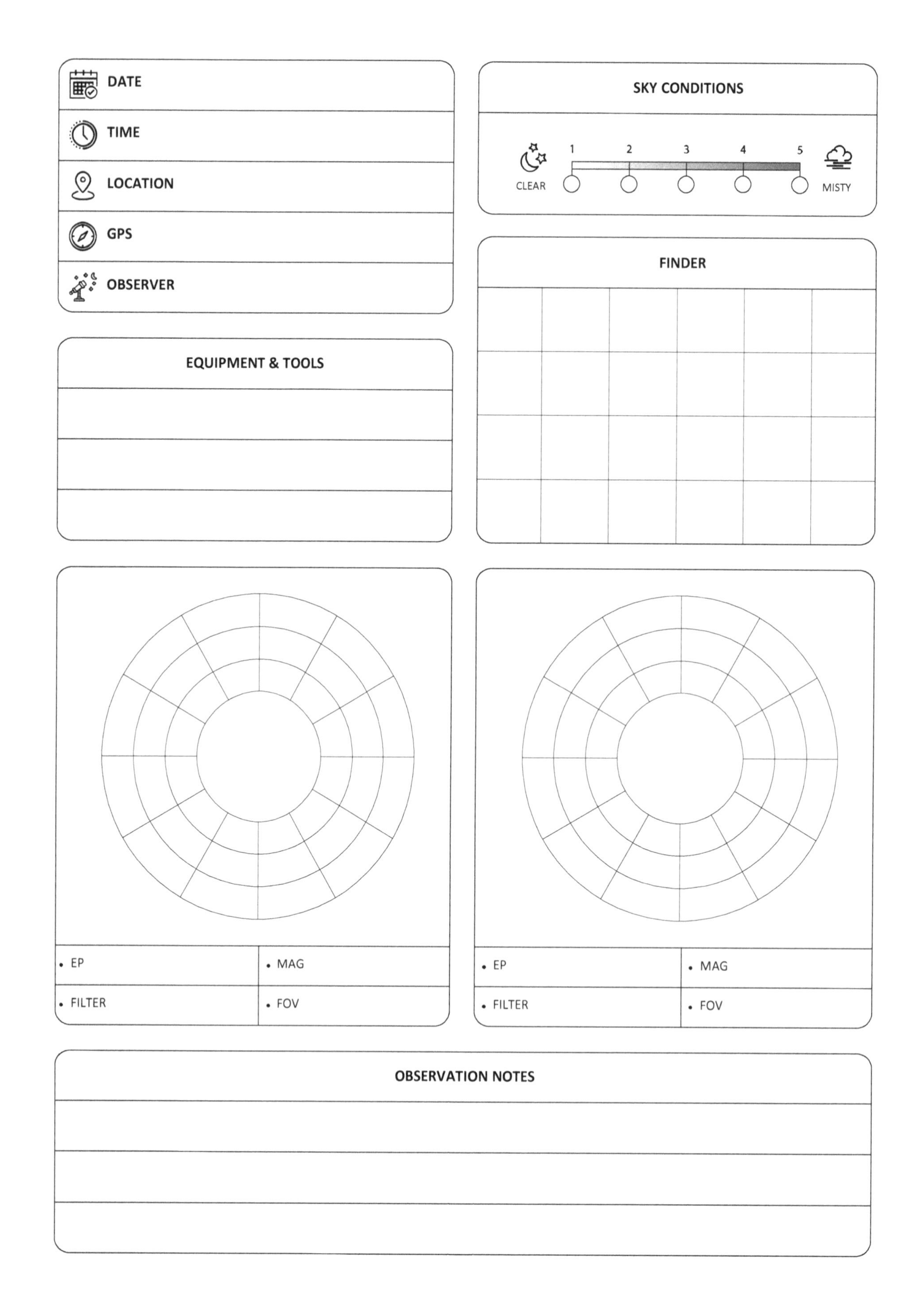
DATE
TIME
LOCATION
GPS
OBSERVER
SKY CONDITIONS
1
2
3
4
5
CLEAR
MISTY
EQUIPMENT & TOOLS
FINDER
EP
MAG
FILTER
FOV
EP
MAG
FILTER
FOV
OBSERVATION NOTES

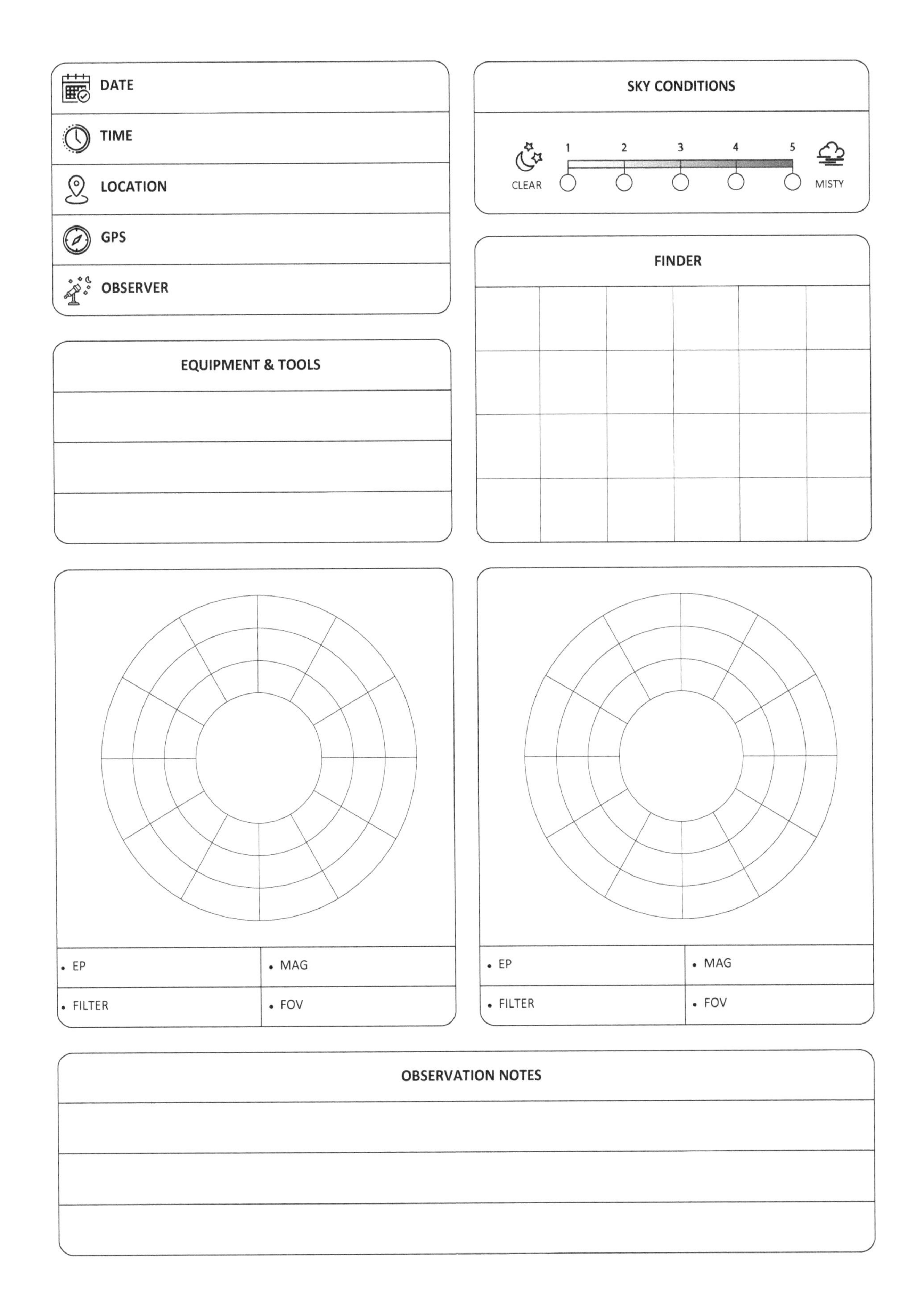

DATE

TIME

LOCATION

GPS

OBSERVER

SKY CONDITIONS

CLEAR 1 2 3 4 5 MISTY

FINDER

EQUIPMENT & TOOLS

- EP
- MAG
- FILTER
- FOV

- EP
- MAG
- FILTER
- FOV

OBSERVATION NOTES

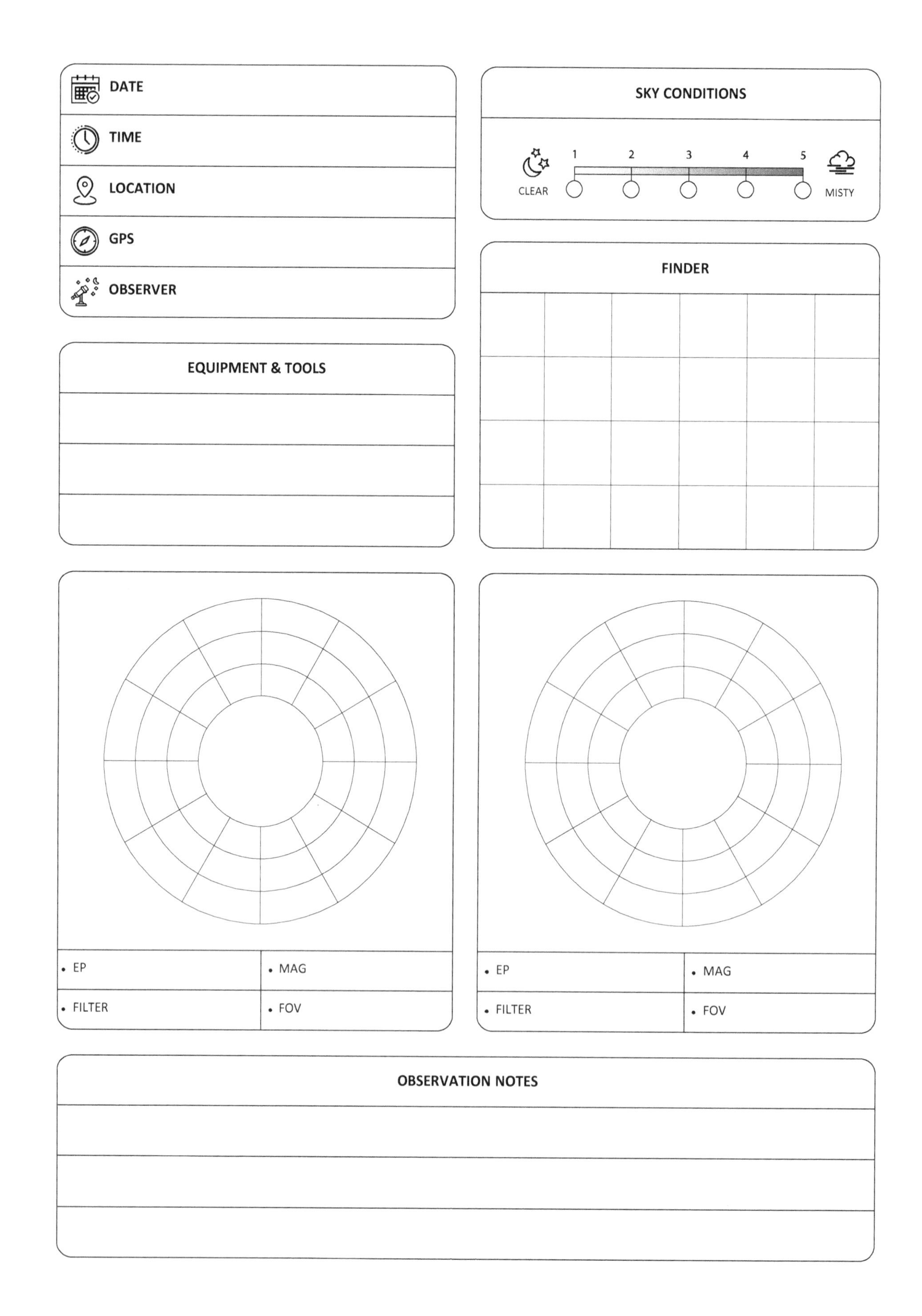

DATE

TIME

LOCATION

GPS

OBSERVER

SKY CONDITIONS

CLEAR 1 2 3 4 5 MISTY

EQUIPMENT & TOOLS

FINDER

- EP
- MAG
- FILTER
- FOV

- EP
- MAG
- FILTER
- FOV

OBSERVATION NOTES

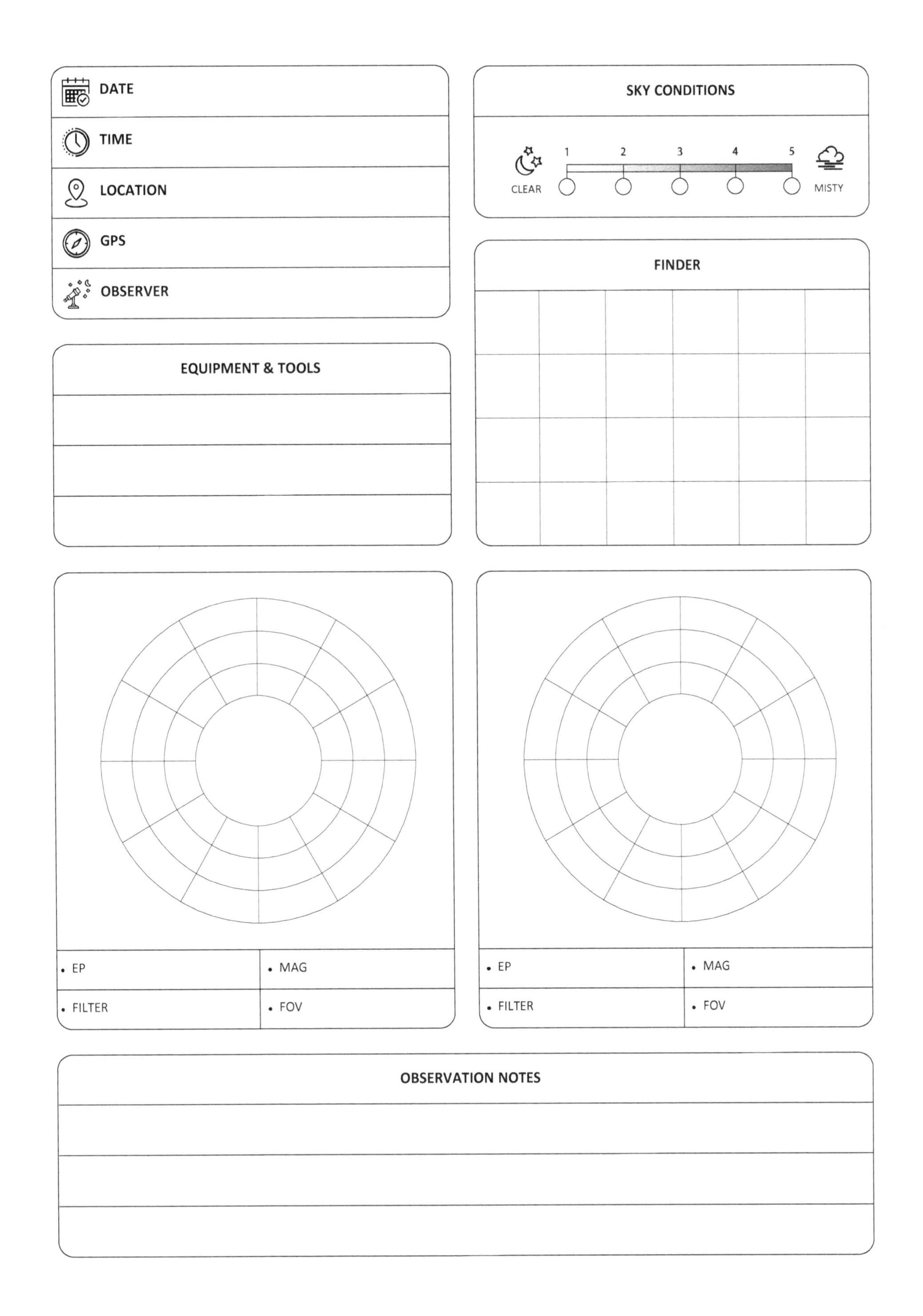

DATE

TIME

LOCATION

GPS

OBSERVER

SKY CONDITIONS

CLEAR 1 2 3 4 5 MISTY

EQUIPMENT & TOOLS

FINDER

EP	MAG
FILTER	FOV

EP	MAG
FILTER	FOV

OBSERVATION NOTES

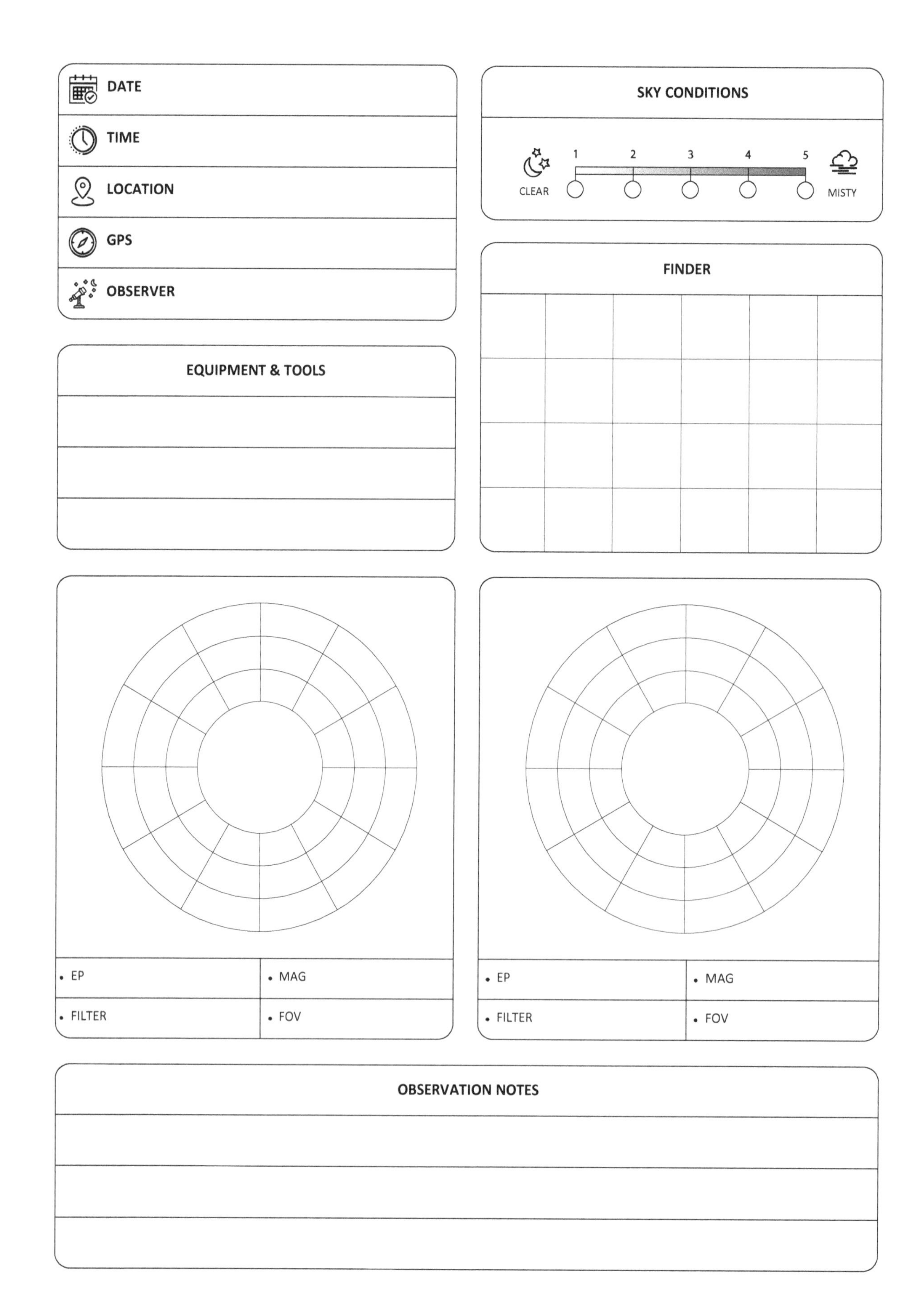
DATE
TIME
LOCATION
GPS
OBSERVER
SKY CONDITIONS
1
2
3
4
5
CLEAR
MISTY
EQUIPMENT & TOOLS
FINDER
EP
MAG
FILTER
FOV
EP
MAG
FILTER
FOV
OBSERVATION NOTES

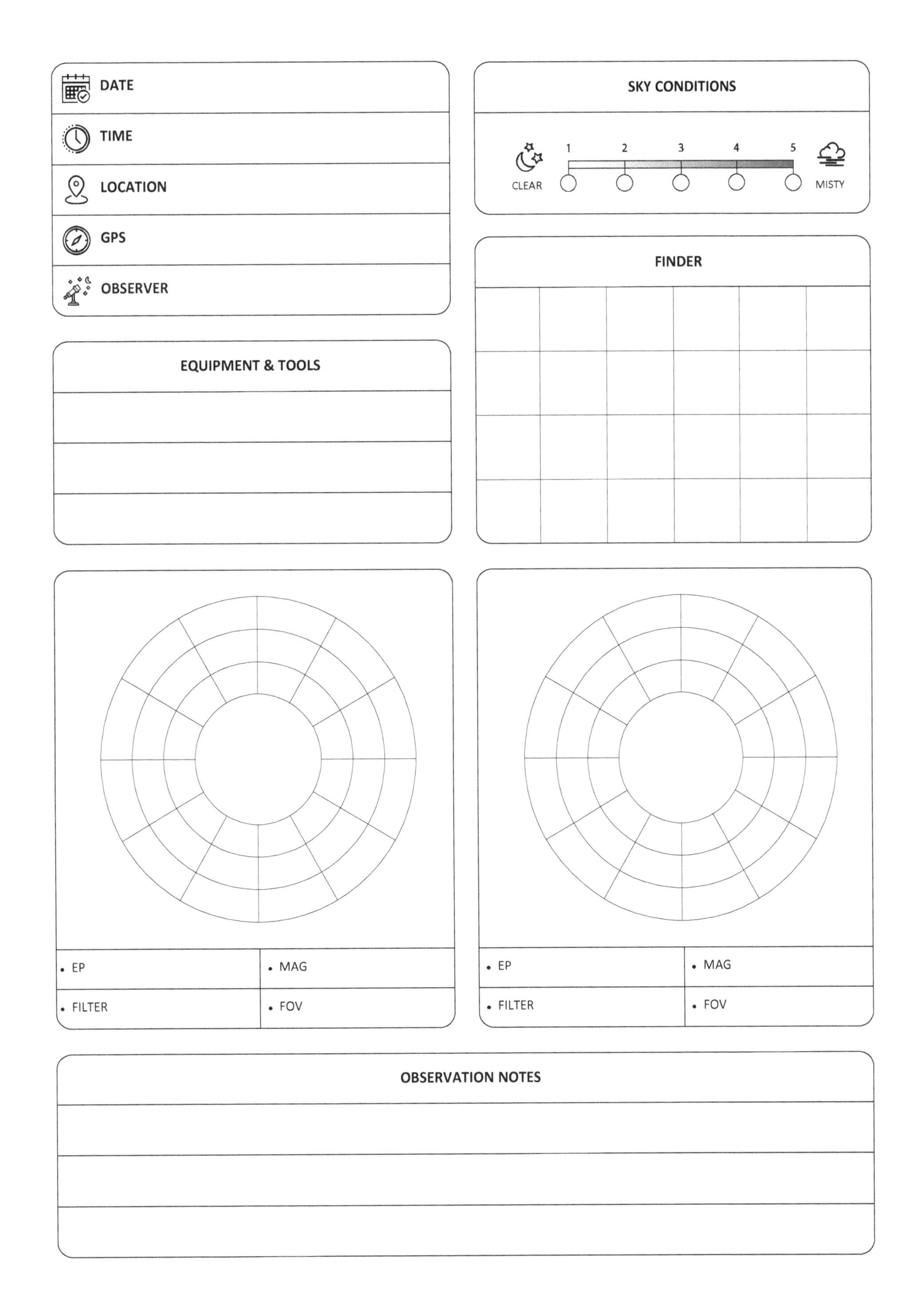
DATE
TIME
LOCATION
GPS
OBSERVER
SKY CONDITIONS
1
2
3
4
5
CLEAR
MISTY
FINDER
EQUIPMENT & TOOLS
EP
MAG
FILTER
FOV
EP
MAG
FILTER
FOV
OBSERVATION NOTES

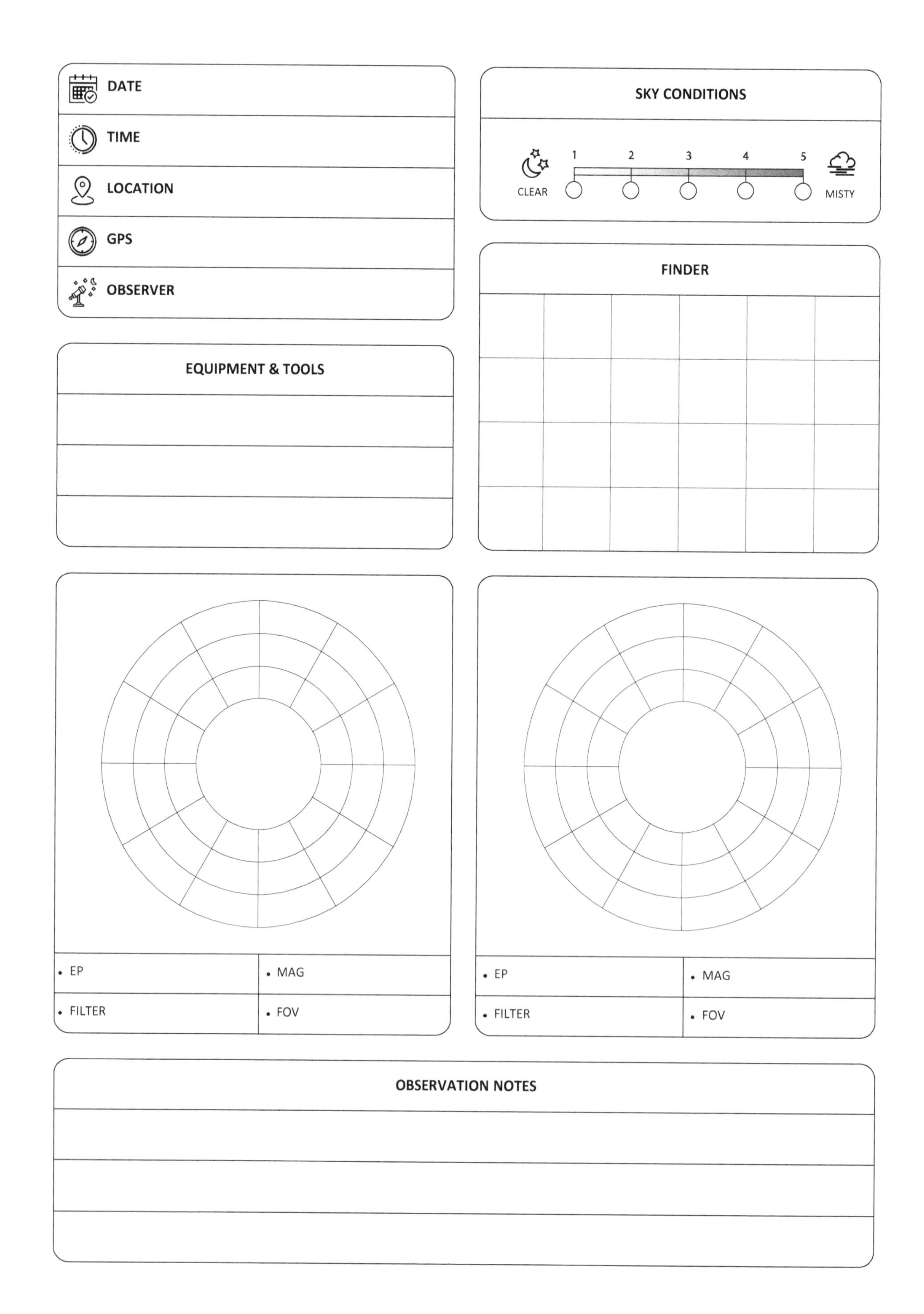
DATE
TIME
LOCATION
GPS
OBSERVER
SKY CONDITIONS
1
2
3
4
5
CLEAR
MISTY
FINDER
EQUIPMENT & TOOLS
EP
MAG
FILTER
FOV
EP
MAG
FILTER
FOV
OBSERVATION NOTES

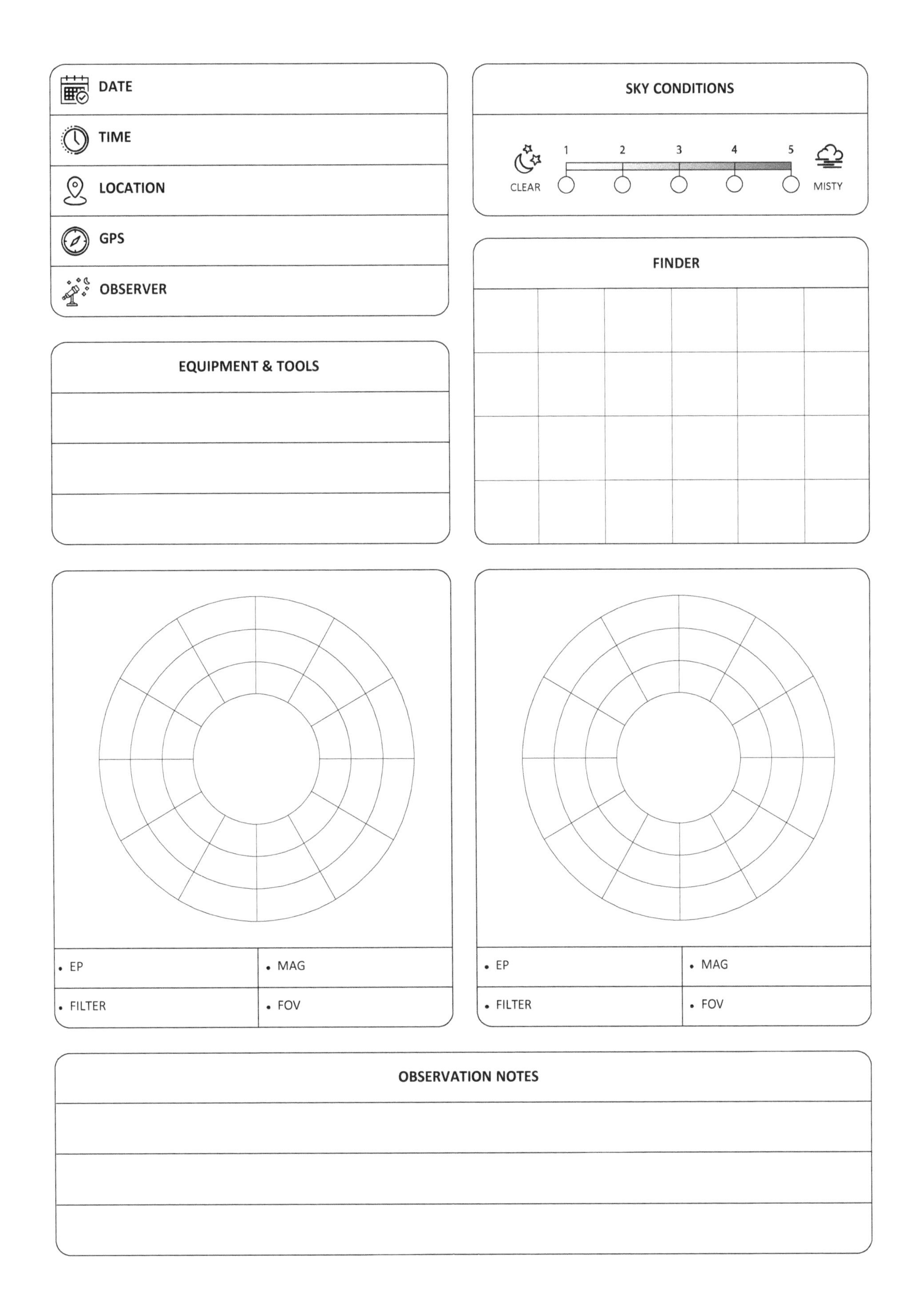
DATE
TIME
LOCATION
GPS
OBSERVER
SKY CONDITIONS
1
2
3
4
5
CLEAR
MISTY
FINDER
EQUIPMENT & TOOLS
EP
MAG
FILTER
FOV
EP
MAG
FILTER
FOV
OBSERVATION NOTES

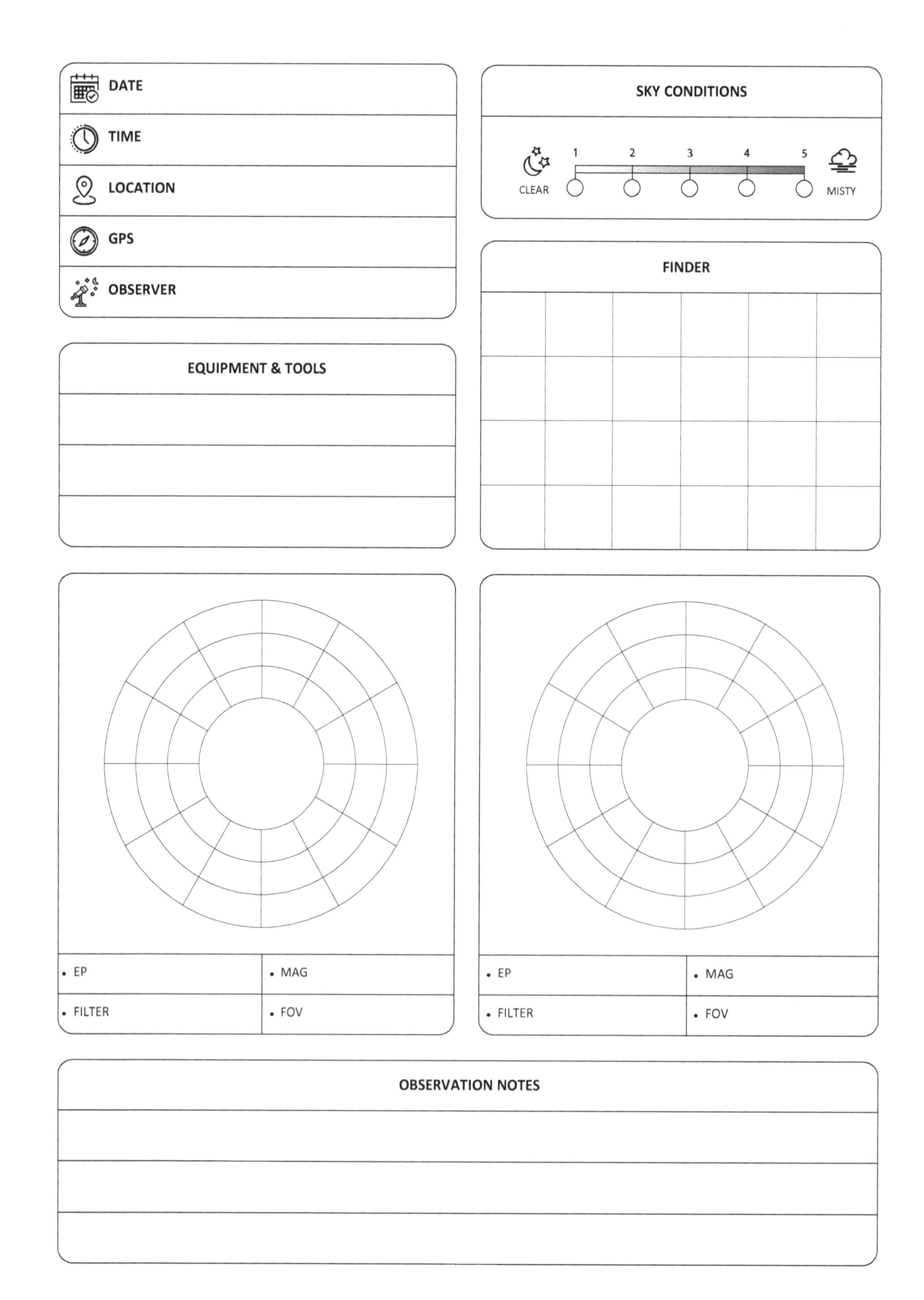
DATE
TIME
LOCATION
GPS
OBSERVER
SKY CONDITIONS
1
2
3
4
5
CLEAR
MISTY
FINDER
EQUIPMENT & TOOLS
EP
MAG
FILTER
FOV
EP
MAG
FILTER
FOV
OBSERVATION NOTES

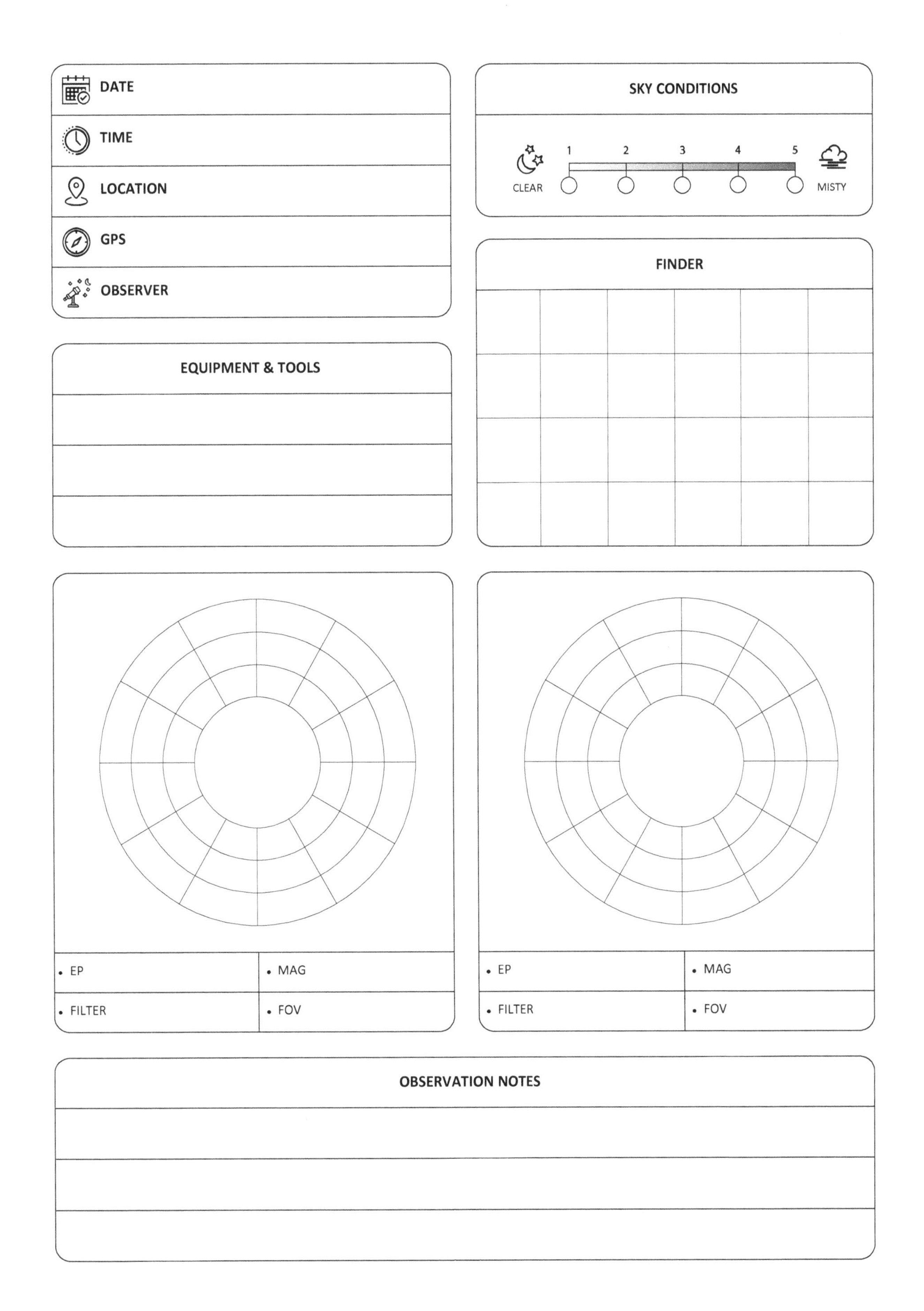
DATE
TIME
LOCATION
GPS
OBSERVER
SKY CONDITIONS
1
2
3
4
5
CLEAR
MISTY
FINDER
EQUIPMENT & TOOLS
EP
MAG
FILTER
FOV
EP
MAG
FILTER
FOV
OBSERVATION NOTES

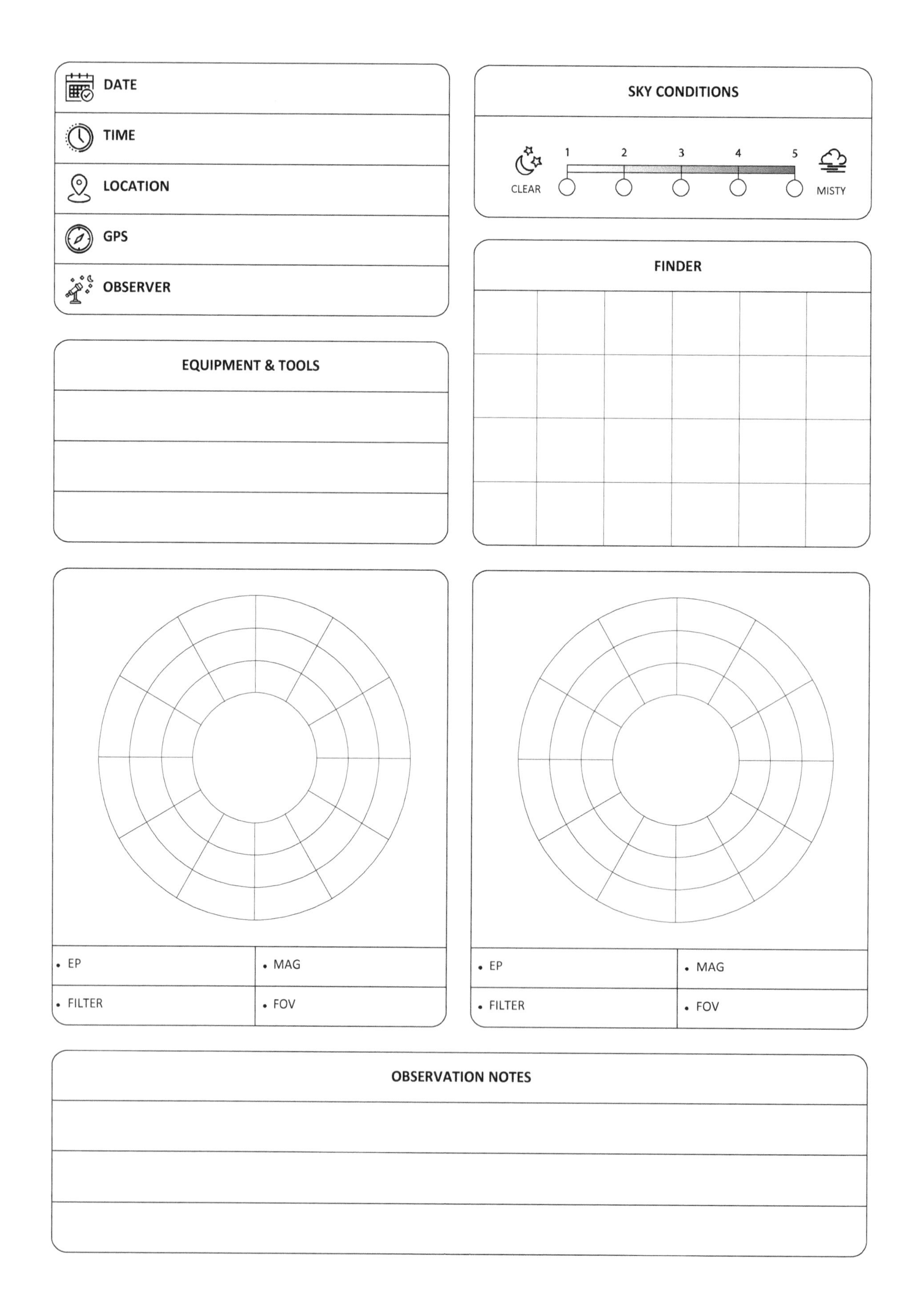
DATE
TIME
LOCATION
GPS
OBSERVER
SKY CONDITIONS
1
2
3
4
5
CLEAR
MISTY
EQUIPMENT & TOOLS
FINDER
• EP
• MAG
• FILTER
• FOV
• EP
• MAG
• FILTER
• FOV
OBSERVATION NOTES

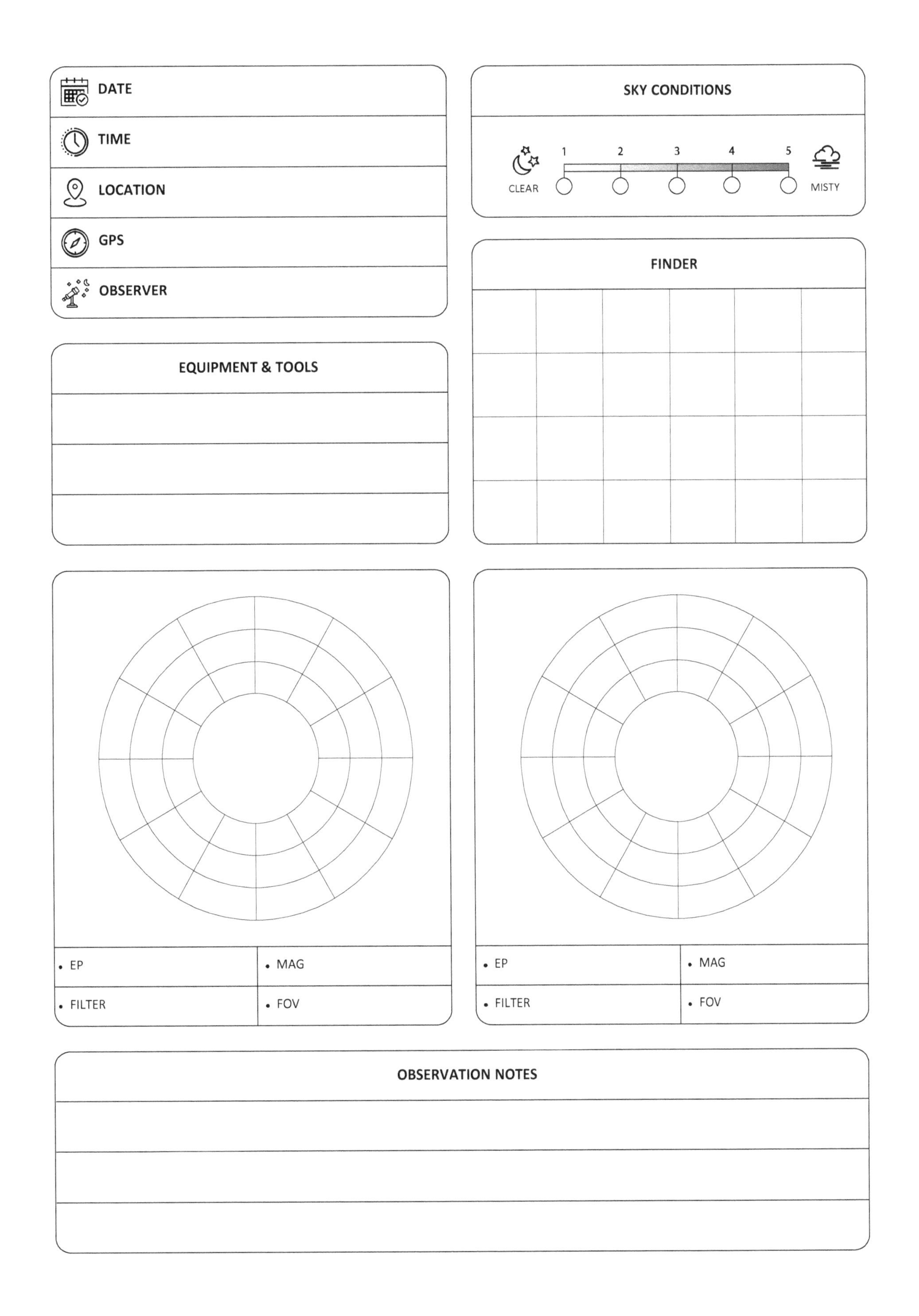
DATE
TIME
LOCATION
GPS
OBSERVER
SKY CONDITIONS
1
2
3
4
5
CLEAR
MISTY
FINDER
EQUIPMENT & TOOLS
EP
MAG
FILTER
FOV
EP
MAG
FILTER
FOV
OBSERVATION NOTES

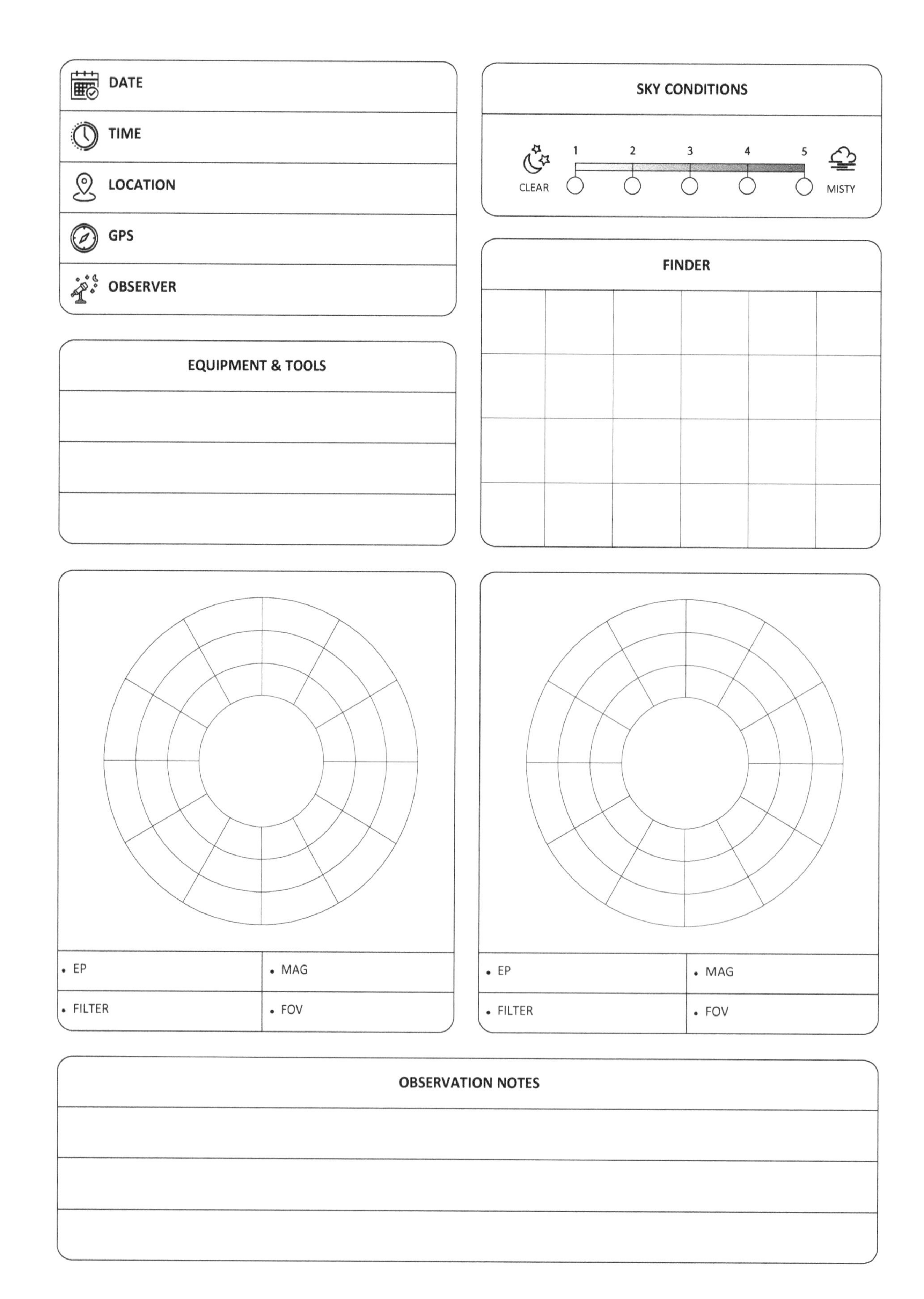
DATE
TIME
LOCATION
GPS
OBSERVER
SKY CONDITIONS
1
2
3
4
5
CLEAR
MISTY
FINDER
EQUIPMENT & TOOLS
EP
MAG
FILTER
FOV
EP
MAG
FILTER
FOV
OBSERVATION NOTES

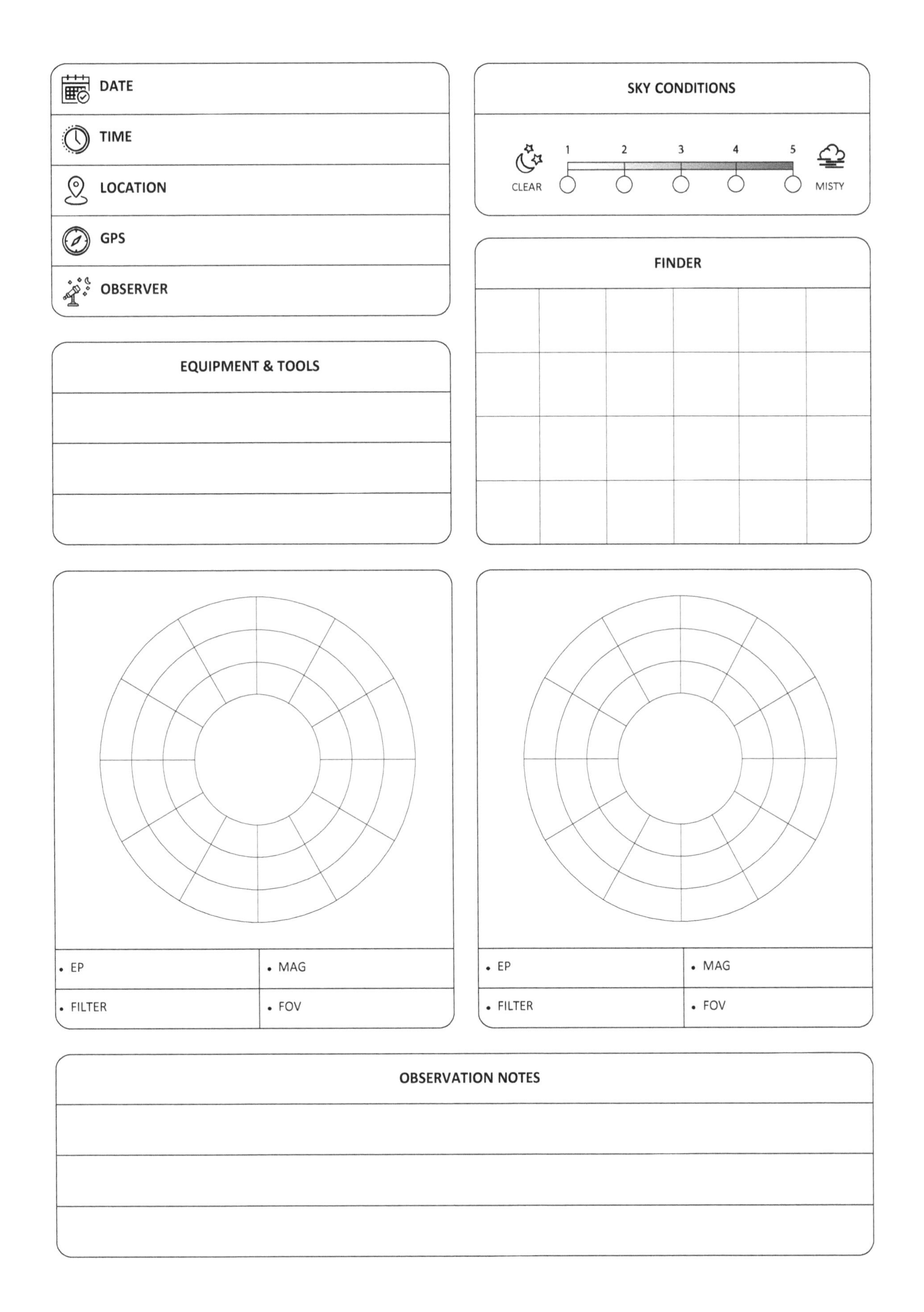
DATE
TIME
LOCATION
GPS
OBSERVER
SKY CONDITIONS
1
2
3
4
5
CLEAR
MISTY
FINDER
EQUIPMENT & TOOLS
EP
MAG
FILTER
FOV
EP
MAG
FILTER
FOV
OBSERVATION NOTES

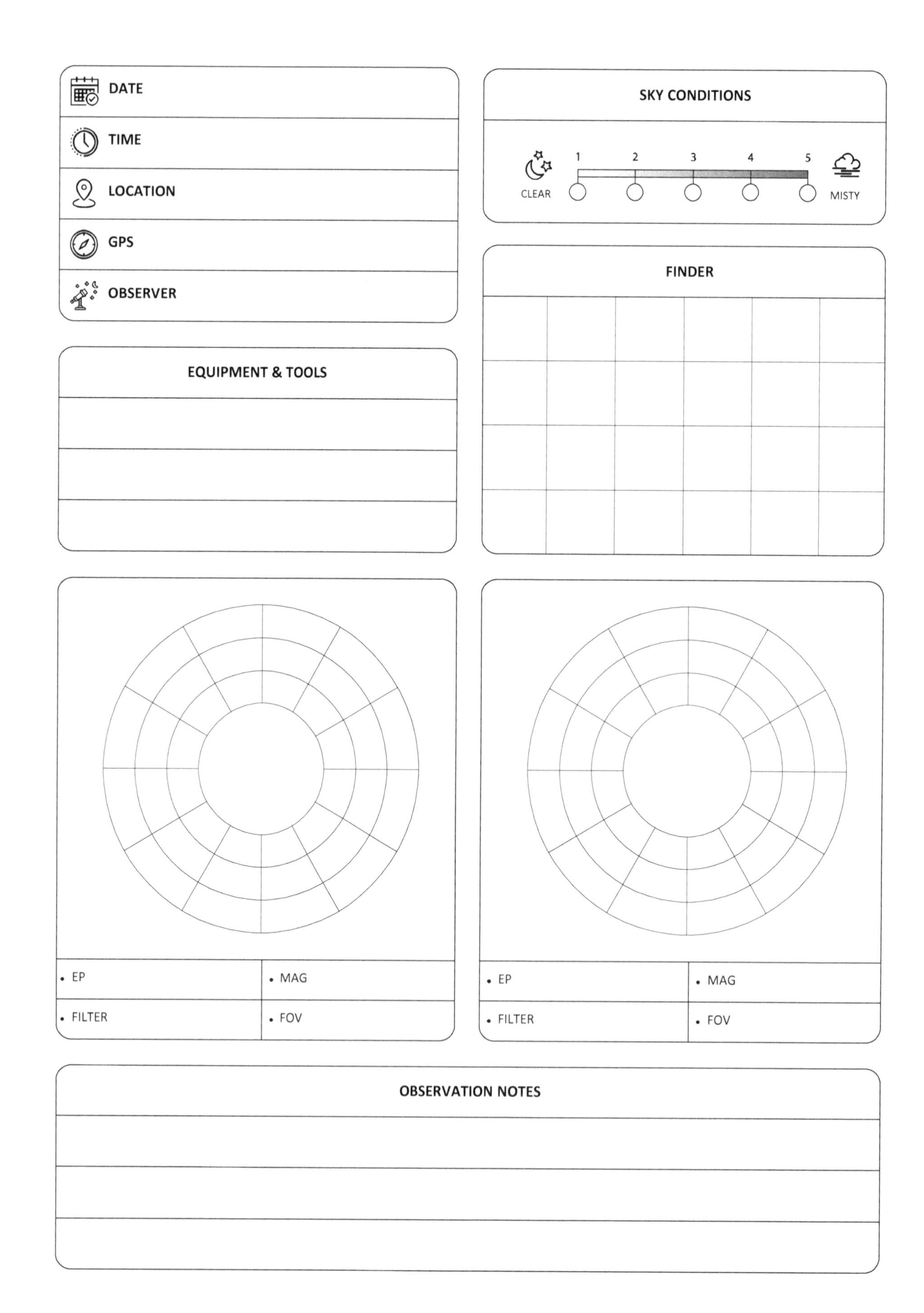
DATE
TIME
LOCATION
GPS
OBSERVER
SKY CONDITIONS
1
2
3
4
5
CLEAR
MISTY
FINDER
EQUIPMENT & TOOLS
• EP
• MAG
• FILTER
• FOV
• EP
• MAG
• FILTER
• FOV
OBSERVATION NOTES

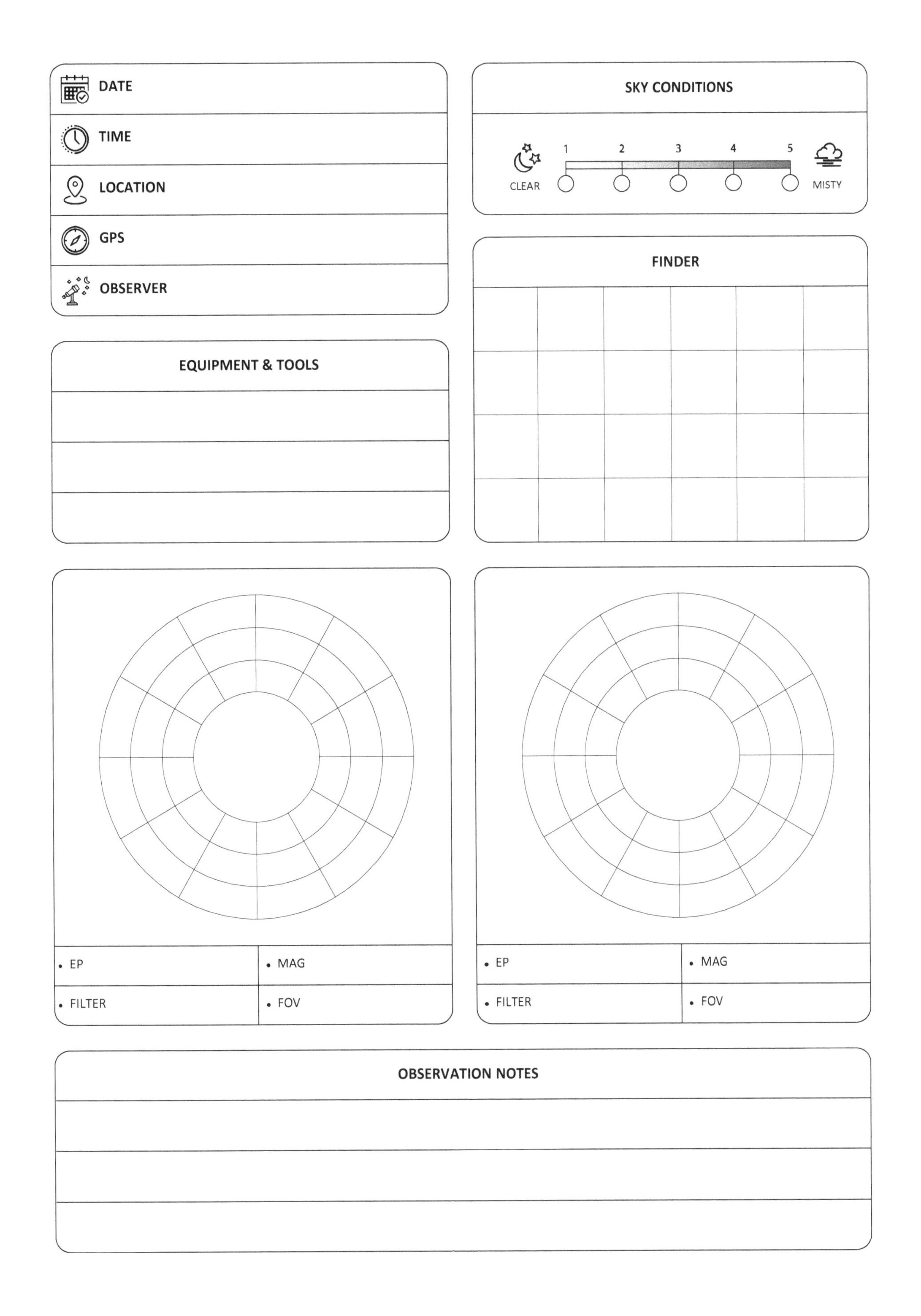
DATE
TIME
LOCATION
GPS
OBSERVER
SKY CONDITIONS
1
2
3
4
5
CLEAR
MISTY
FINDER
EQUIPMENT & TOOLS
EP
MAG
FILTER
FOV
EP
MAG
FILTER
FOV
OBSERVATION NOTES

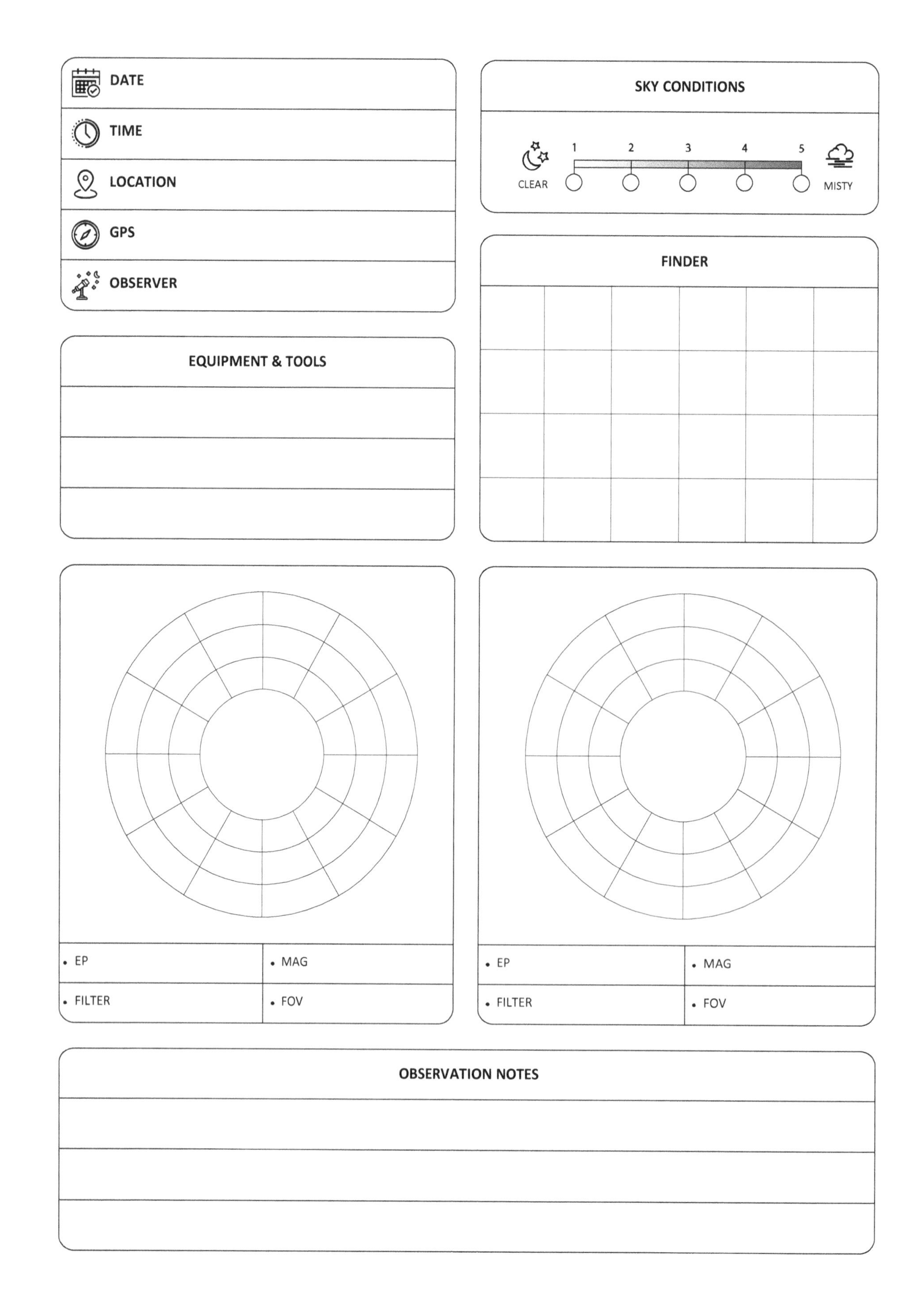

DATE
TIME
LOCATION
GPS
OBSERVER
SKY CONDITIONS
1
2
3
4
5
CLEAR
MISTY
FINDER
EQUIPMENT & TOOLS
EP
MAG
FILTER
FOV
EP
MAG
FILTER
FOV
OBSERVATION NOTES

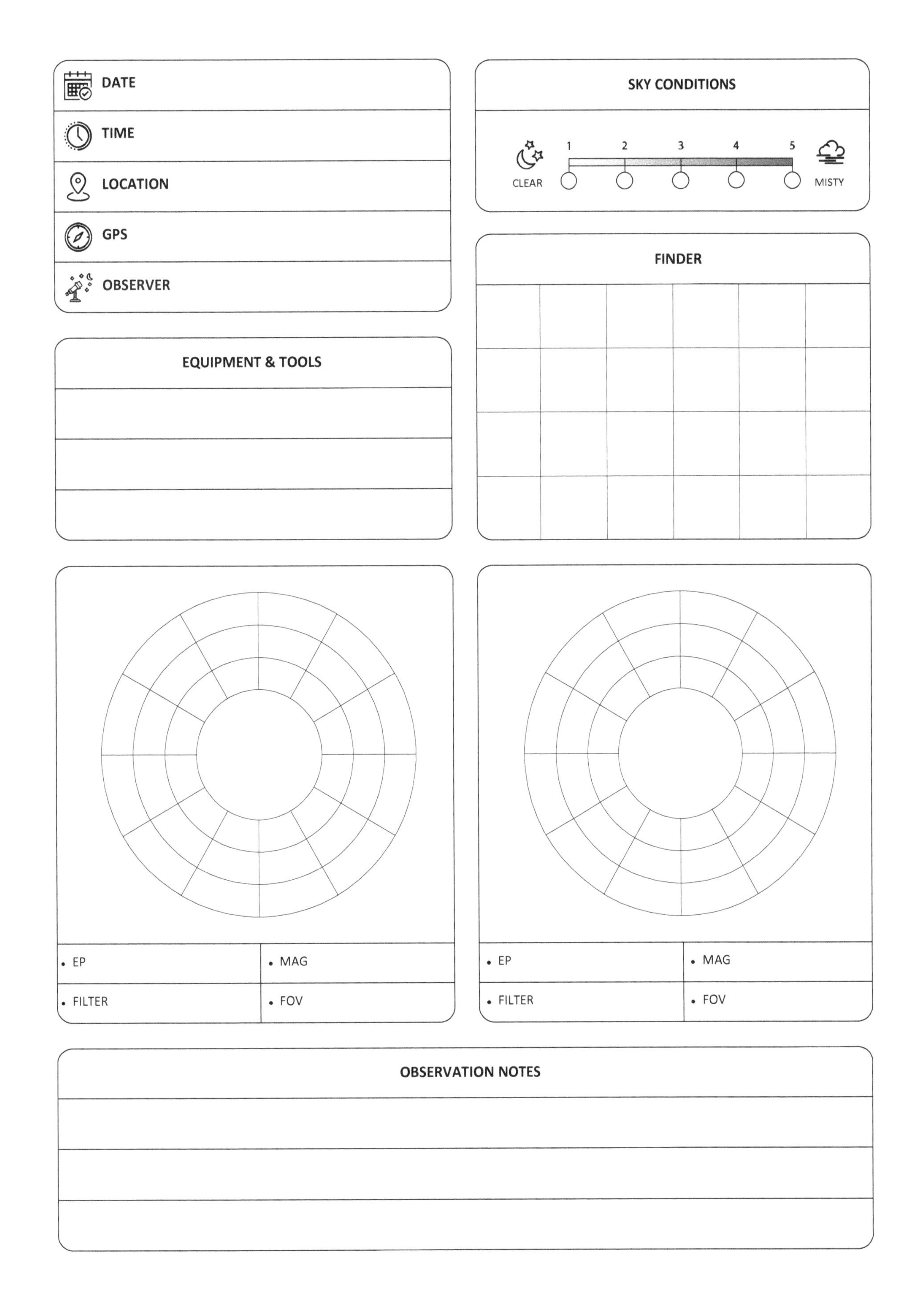
DATE
TIME
LOCATION
GPS
OBSERVER
SKY CONDITIONS
1
2
3
4
5
CLEAR
MISTY
FINDER
EQUIPMENT & TOOLS
EP
MAG
FILTER
FOV
EP
MAG
FILTER
FOV
OBSERVATION NOTES

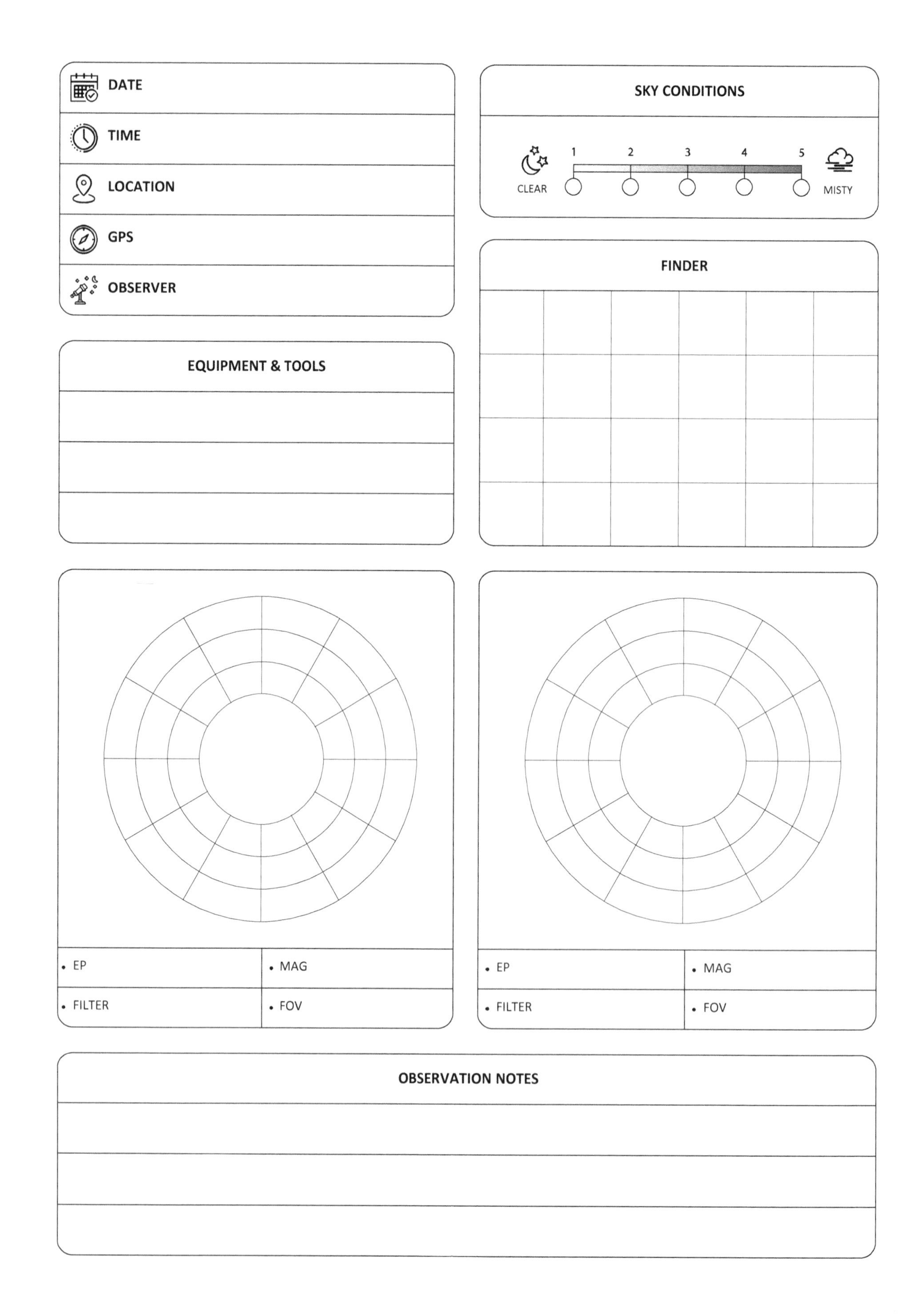
DATE
TIME
LOCATION
GPS
OBSERVER
SKY CONDITIONS
1
2
3
4
5
CLEAR
MISTY
FINDER
EQUIPMENT & TOOLS
EP
MAG
FILTER
FOV
EP
MAG
FILTER
FOV
OBSERVATION NOTES

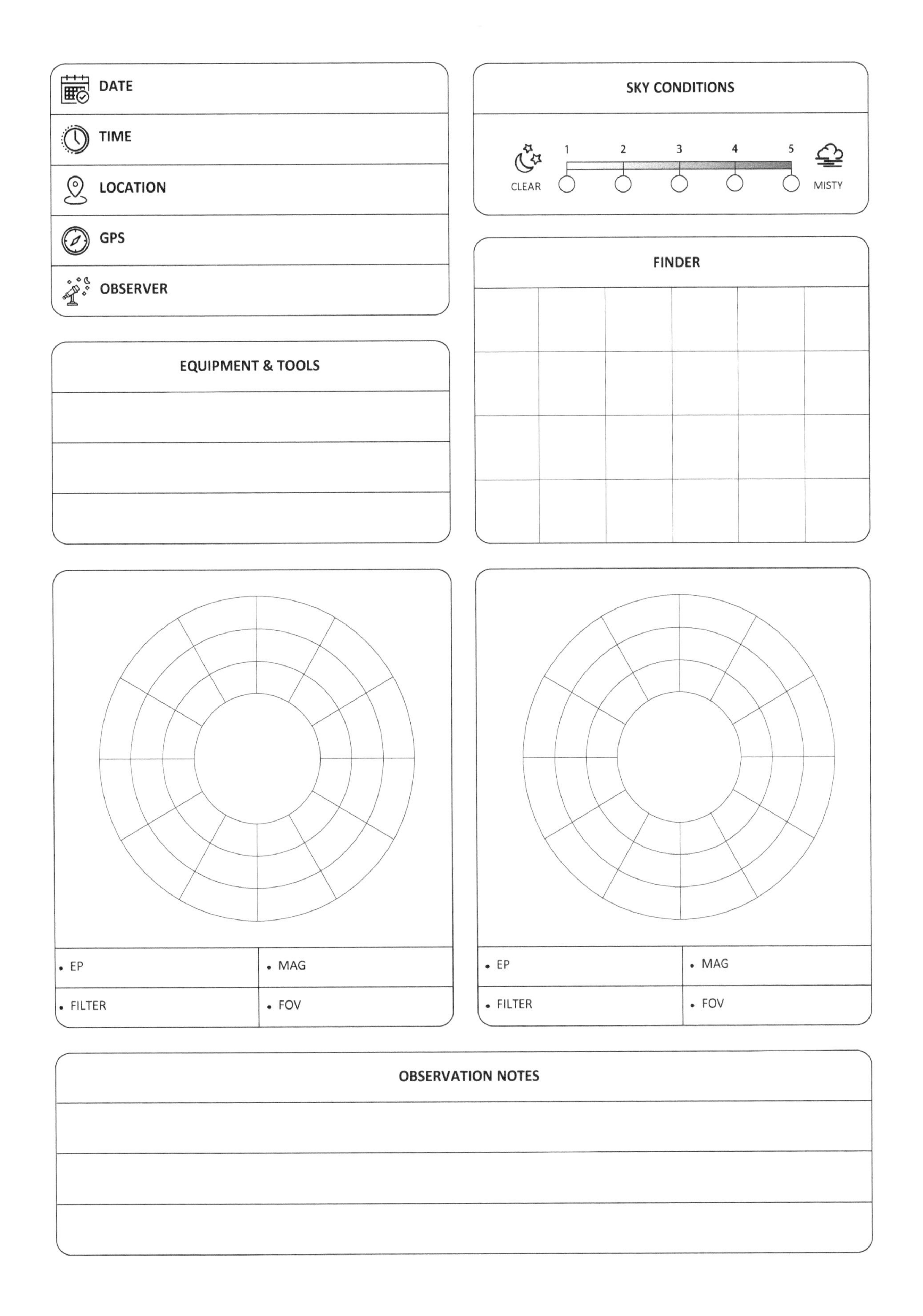
DATE
TIME
LOCATION
GPS
OBSERVER
SKY CONDITIONS
1
2
3
4
5
CLEAR
MISTY
FINDER
EQUIPMENT & TOOLS
EP
MAG
FILTER
FOV
EP
MAG
FILTER
FOV
OBSERVATION NOTES

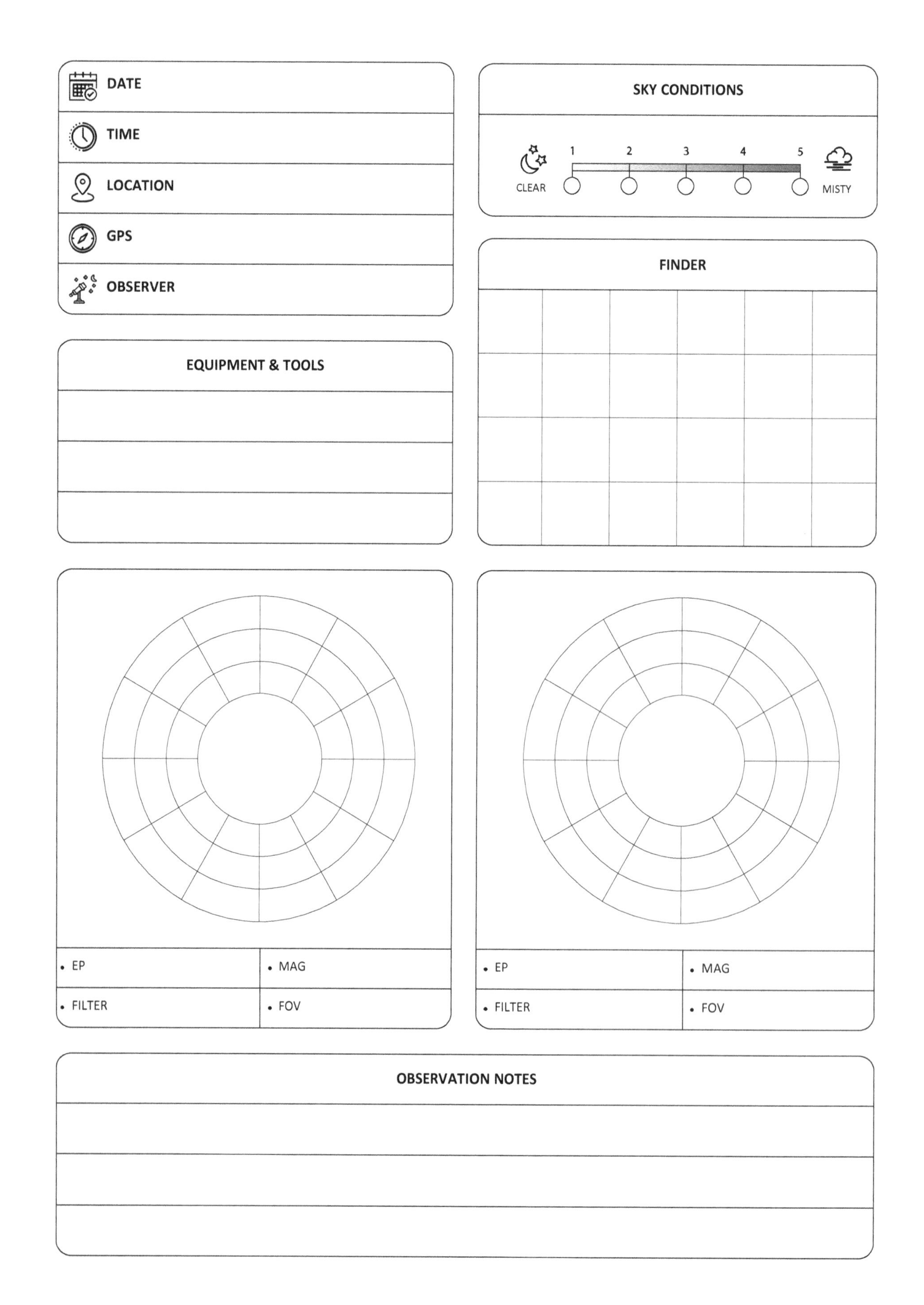
DATE
TIME
LOCATION
GPS
OBSERVER
SKY CONDITIONS
1
2
3
4
5
CLEAR
MISTY
FINDER
EQUIPMENT & TOOLS
EP
MAG
FILTER
FOV
EP
MAG
FILTER
FOV
OBSERVATION NOTES

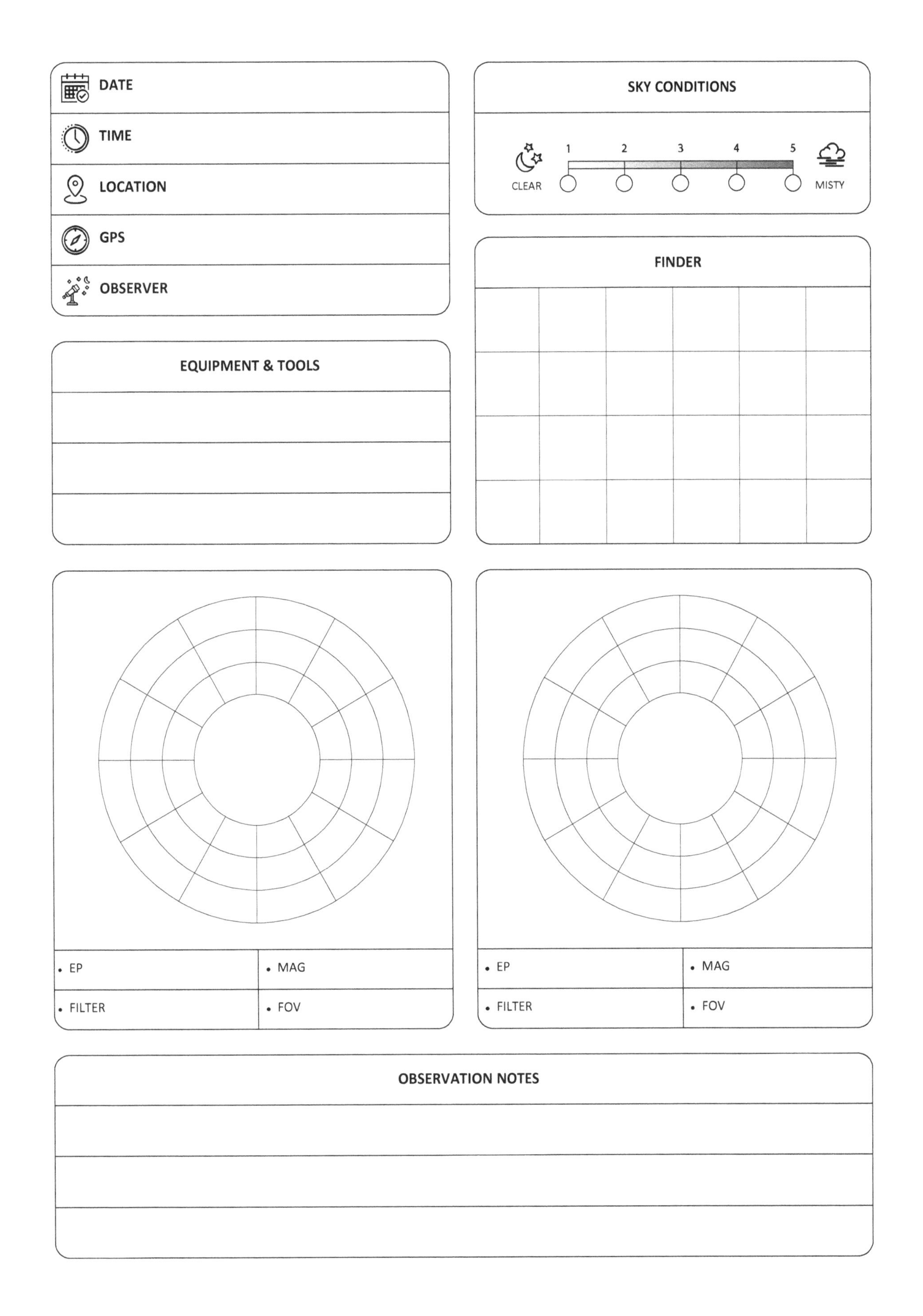
DATE
TIME
LOCATION
GPS
OBSERVER
SKY CONDITIONS
1
2
3
4
5
CLEAR
MISTY
FINDER
EQUIPMENT & TOOLS
EP
MAG
FILTER
FOV
EP
MAG
FILTER
FOV
OBSERVATION NOTES

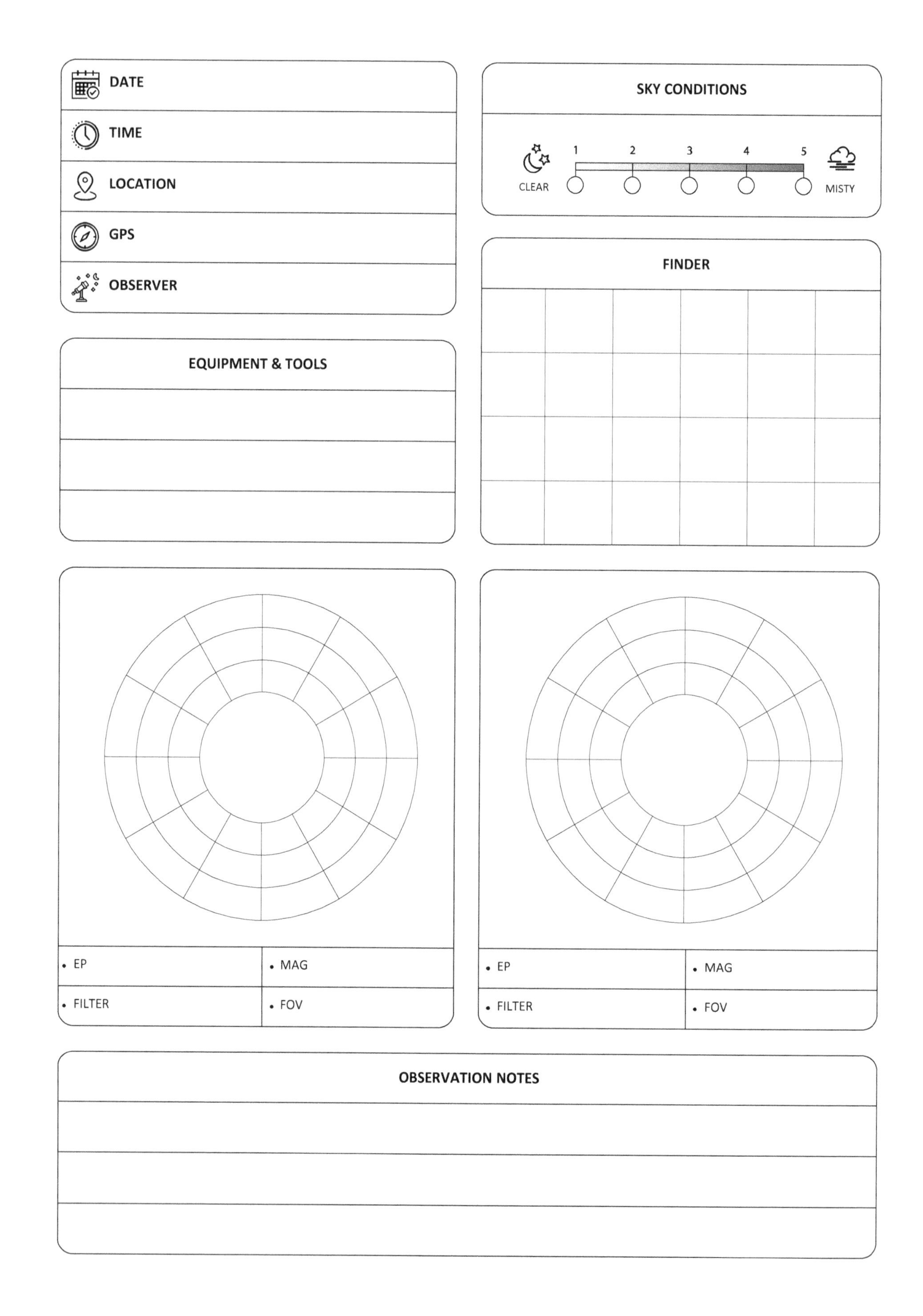
DATE
TIME
LOCATION
GPS
OBSERVER
SKY CONDITIONS
1
2
3
4
5
CLEAR
MISTY
FINDER
EQUIPMENT & TOOLS
EP
MAG
FILTER
FOV
EP
MAG
FILTER
FOV
OBSERVATION NOTES

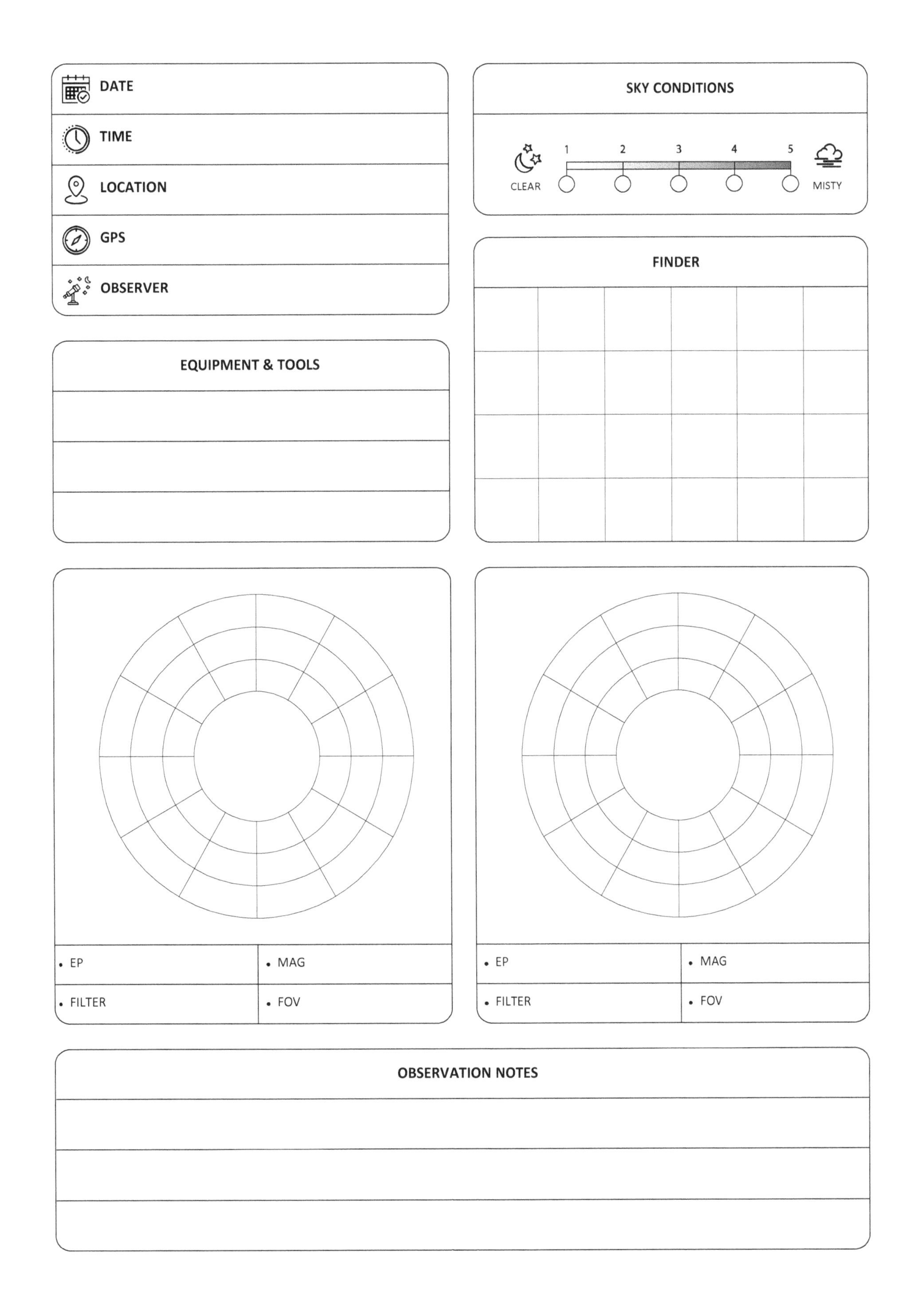
DATE
TIME
LOCATION
GPS
OBSERVER
SKY CONDITIONS
1
2
3
4
5
CLEAR
MISTY
FINDER
EQUIPMENT & TOOLS
EP
MAG
FILTER
FOV
EP
MAG
FILTER
FOV
OBSERVATION NOTES

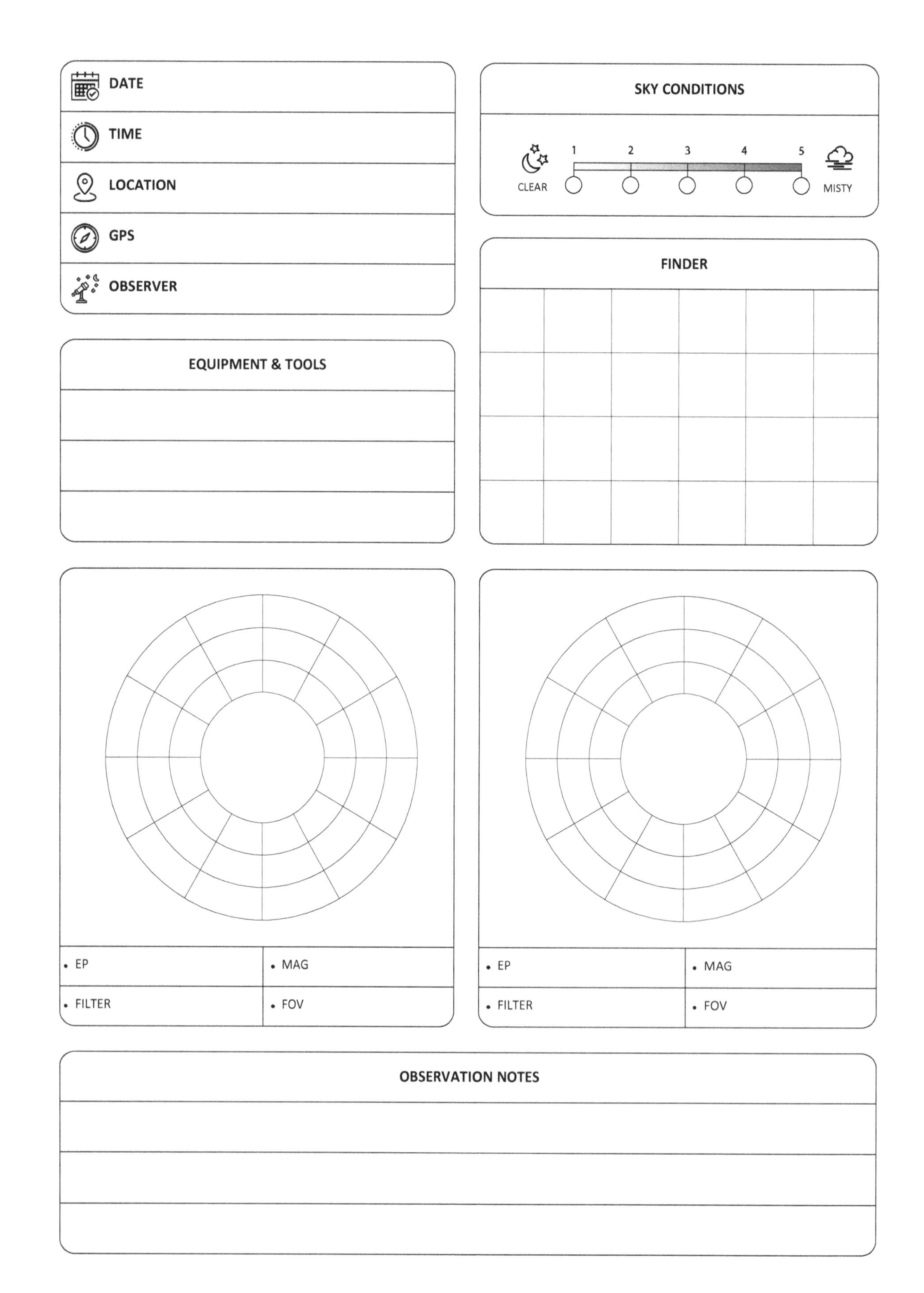
DATE
TIME
LOCATION
GPS
OBSERVER
SKY CONDITIONS
1
2
3
4
5
CLEAR
MISTY
EQUIPMENT & TOOLS
FINDER
EP
MAG
FILTER
FOV
EP
MAG
FILTER
FOV
OBSERVATION NOTES

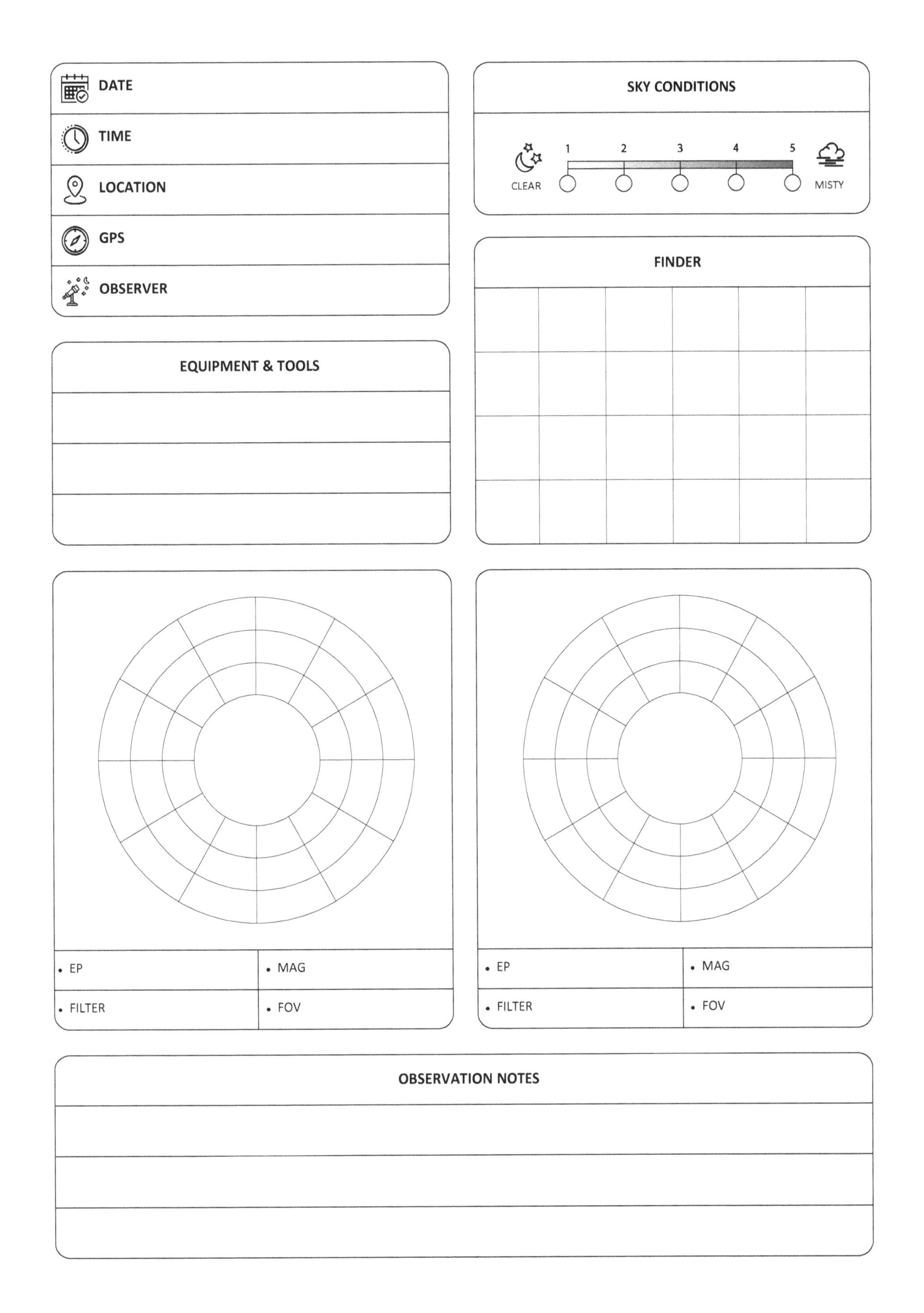
DATE
TIME
LOCATION
GPS
OBSERVER
SKY CONDITIONS
1
2
3
4
5
CLEAR
MISTY
FINDER
EQUIPMENT & TOOLS
EP
MAG
FILTER
FOV
EP
MAG
FILTER
FOV
OBSERVATION NOTES

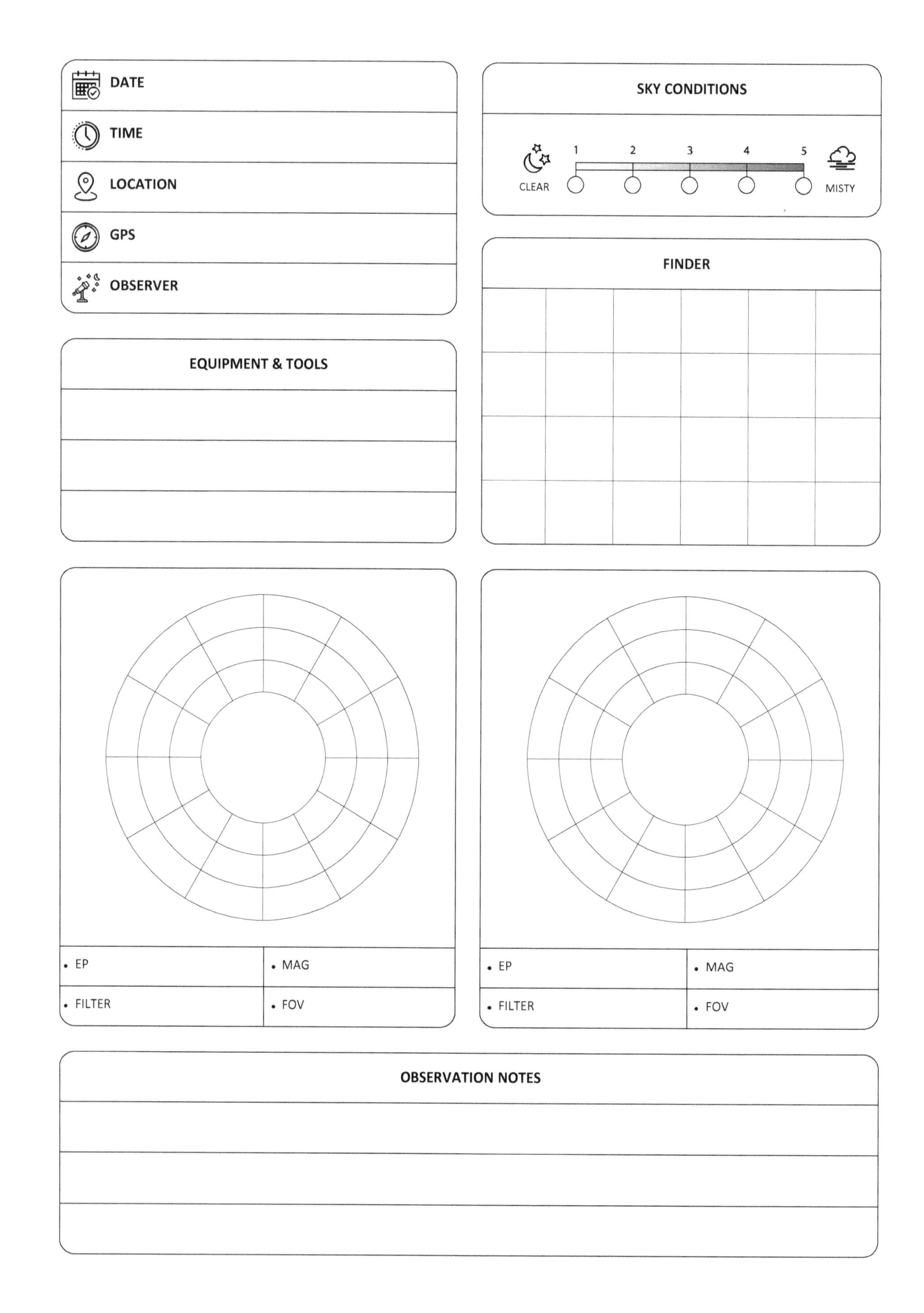
DATE
TIME
LOCATION
GPS
OBSERVER
SKY CONDITIONS
1
2
3
4
5
CLEAR
MISTY
FINDER
EQUIPMENT & TOOLS
EP
MAG
FILTER
FOV
EP
MAG
FILTER
FOV
OBSERVATION NOTES

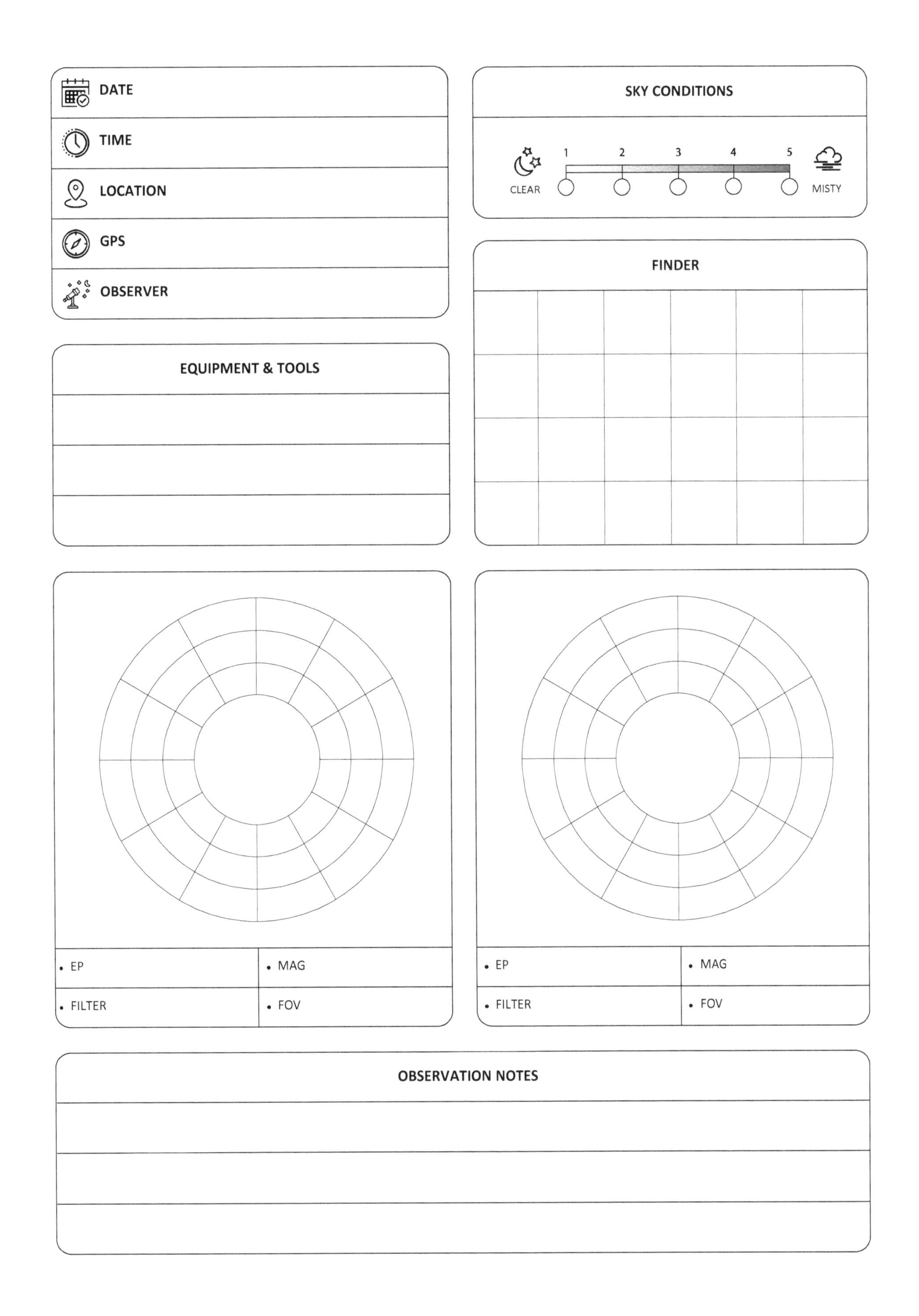

DATE
TIME
LOCATION
GPS
OBSERVER
SKY CONDITIONS
1
2
3
4
5
CLEAR
MISTY
FINDER
EQUIPMENT & TOOLS
EP
MAG
FILTER
FOV
EP
MAG
FILTER
FOV
OBSERVATION NOTES

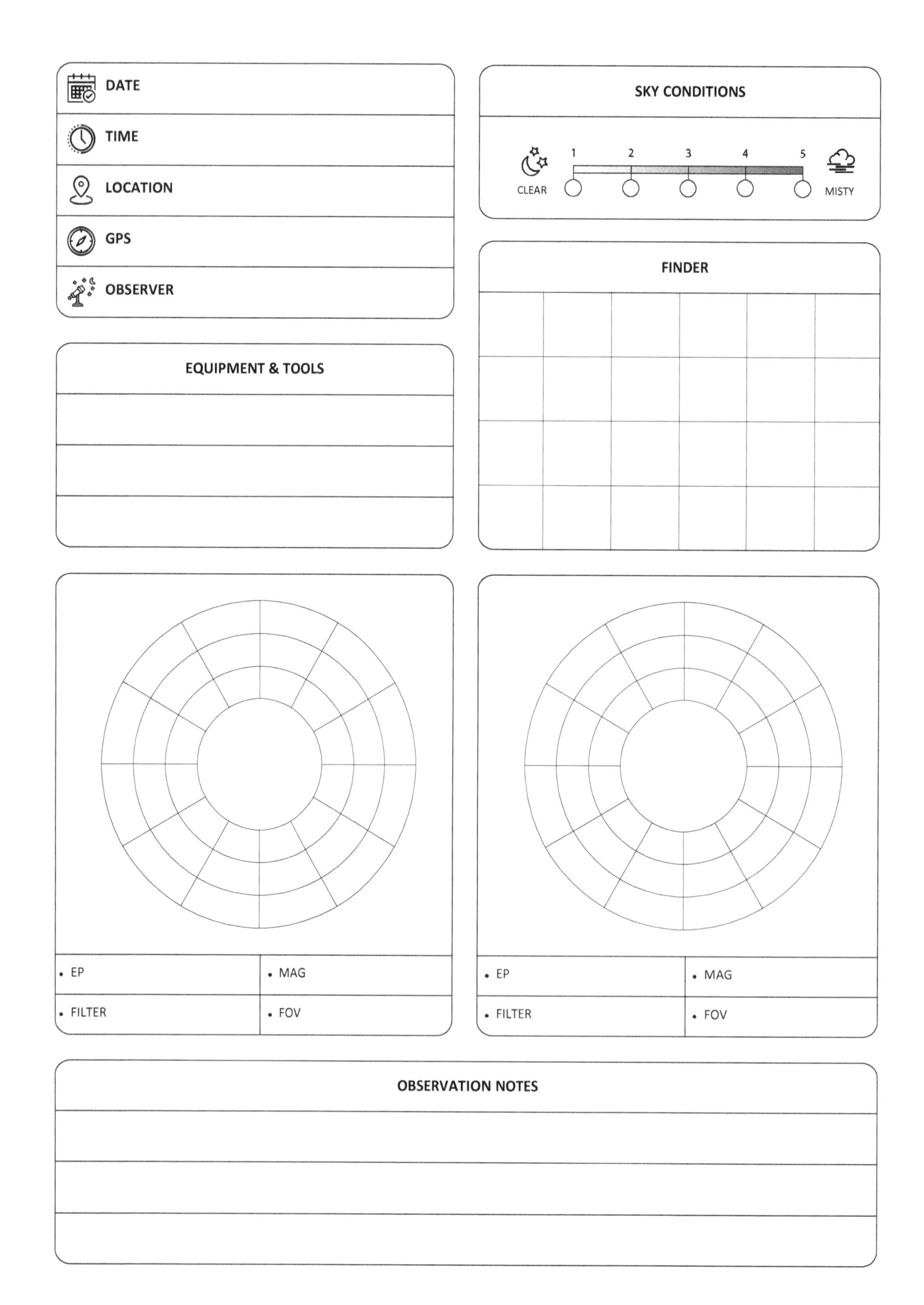
DATE
TIME
LOCATION
GPS
OBSERVER
SKY CONDITIONS
1
2
3
4
5
CLEAR
MISTY
FINDER
EQUIPMENT & TOOLS
EP
MAG
FILTER
FOV
EP
MAG
FILTER
FOV
OBSERVATION NOTES

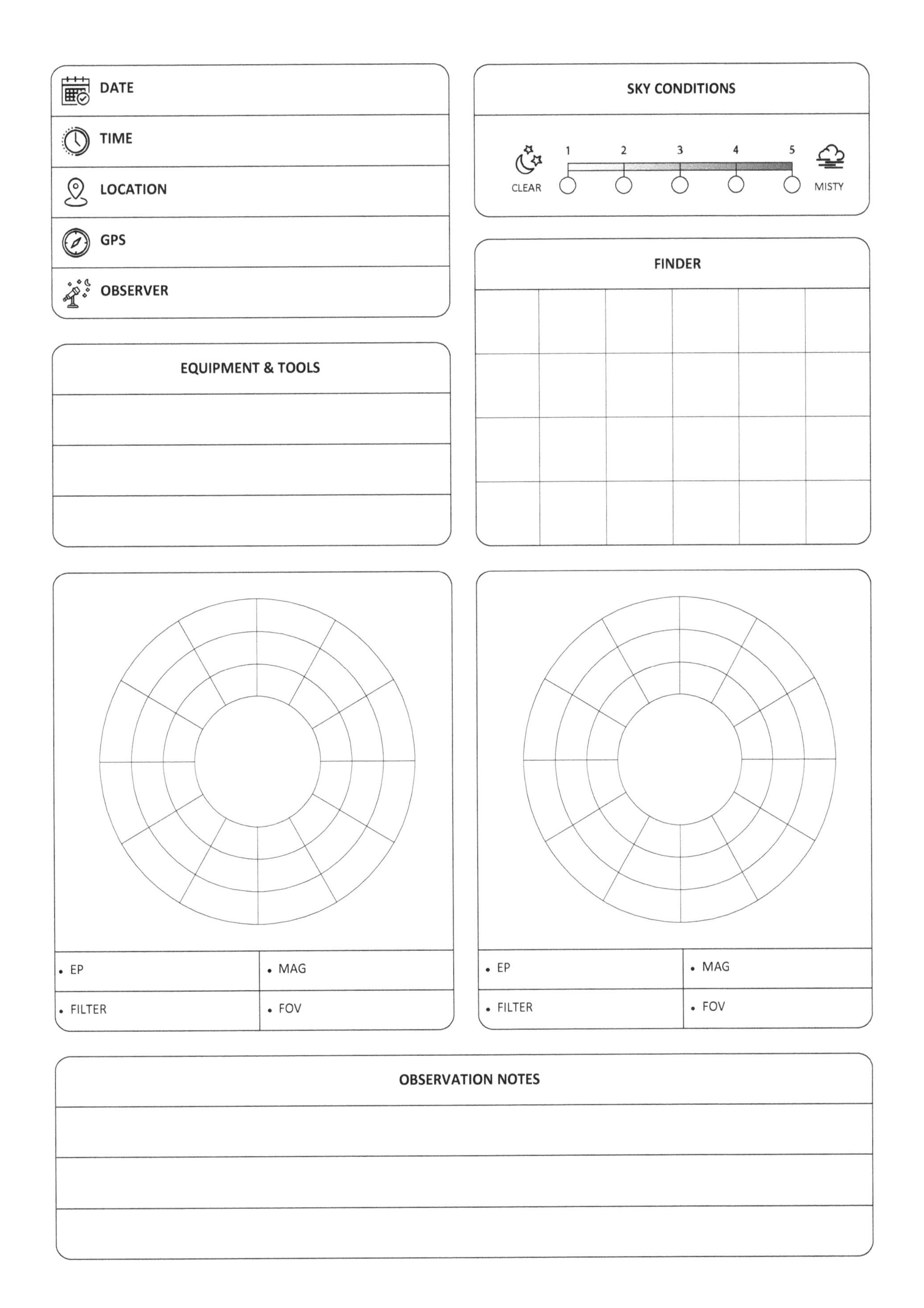
DATE
TIME
LOCATION
GPS
OBSERVER
SKY CONDITIONS
1
2
3
4
5
CLEAR
MISTY
FINDER
EQUIPMENT & TOOLS
EP
MAG
FILTER
FOV
EP
MAG
FILTER
FOV
OBSERVATION NOTES

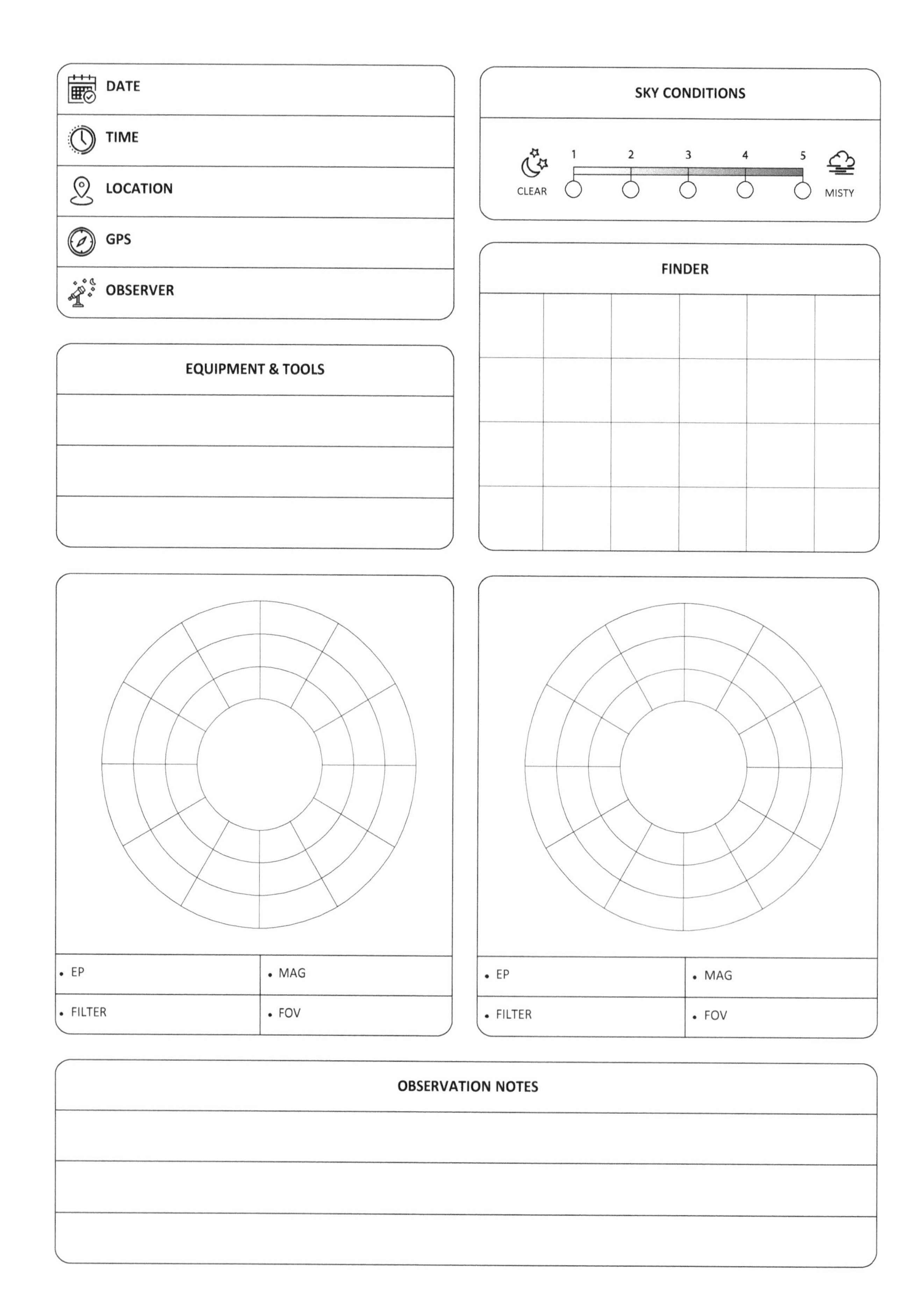
DATE
TIME
LOCATION
GPS
OBSERVER

SKY CONDITIONS

CLEAR 1 2 3 4 5 MISTY

FINDER

EQUIPMENT & TOOLS

- EP
- MAG
- FILTER
- FOV

- EP
- MAG
- FILTER
- FOV

OBSERVATION NOTES

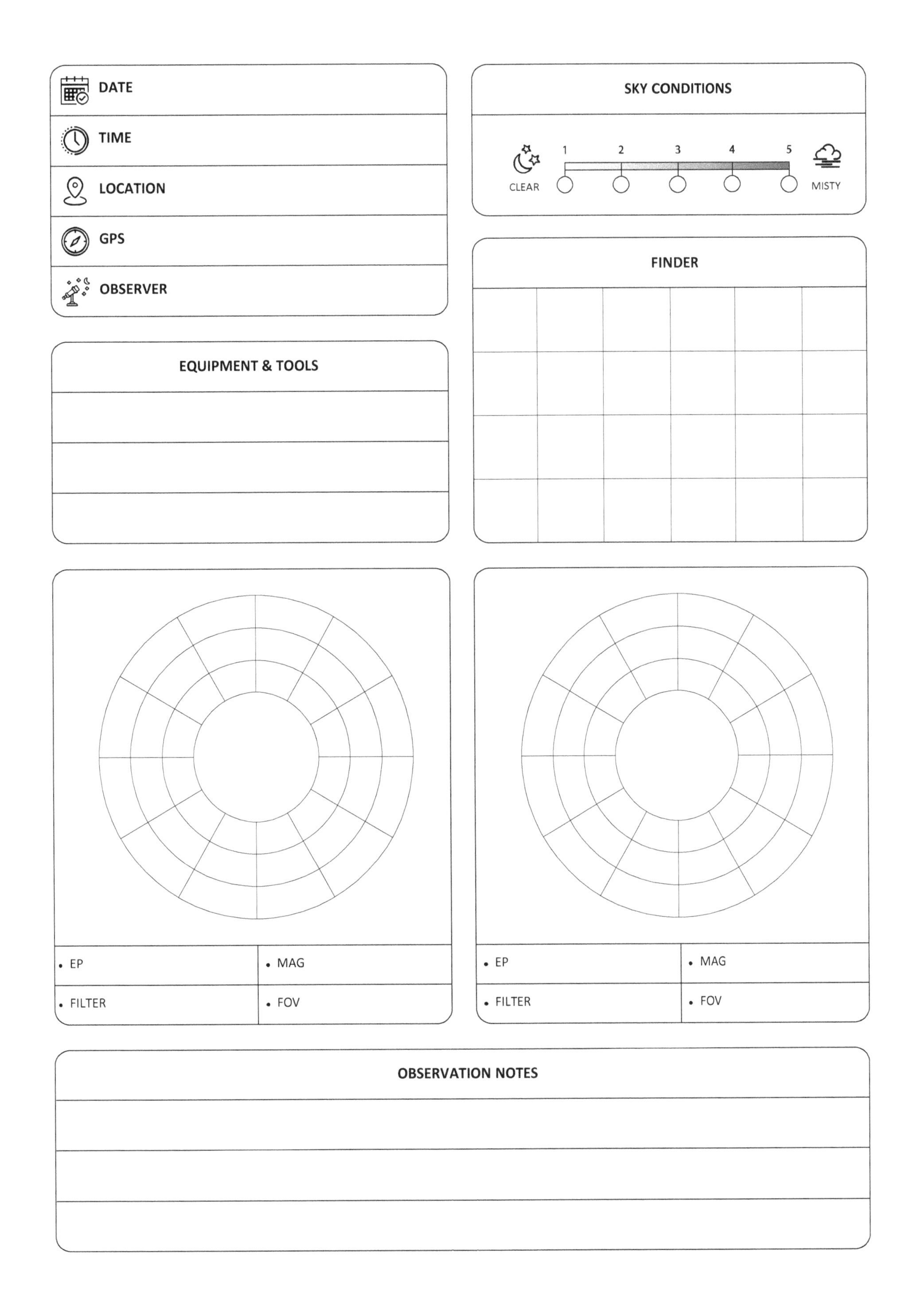
DATE
TIME
LOCATION
GPS
OBSERVER
SKY CONDITIONS
1
2
3
4
5
CLEAR
MISTY
FINDER
EQUIPMENT & TOOLS
EP
MAG
FILTER
FOV
EP
MAG
FILTER
FOV
OBSERVATION NOTES

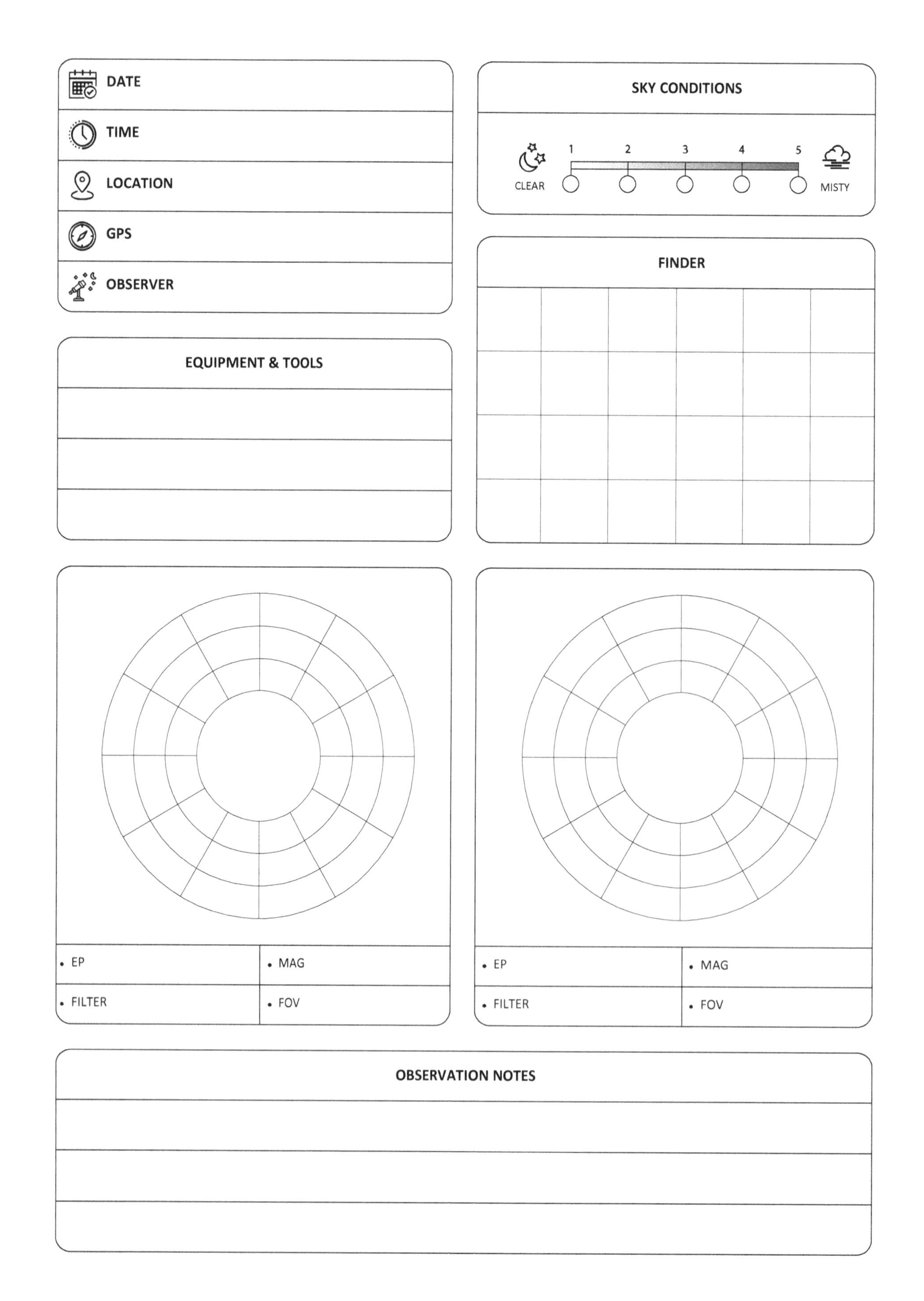
DATE
TIME
LOCATION
GPS
OBSERVER
SKY CONDITIONS
1
2
3
4
5
CLEAR
MISTY
FINDER
EQUIPMENT & TOOLS
EP
MAG
FILTER
FOV
EP
MAG
FILTER
FOV
OBSERVATION NOTES

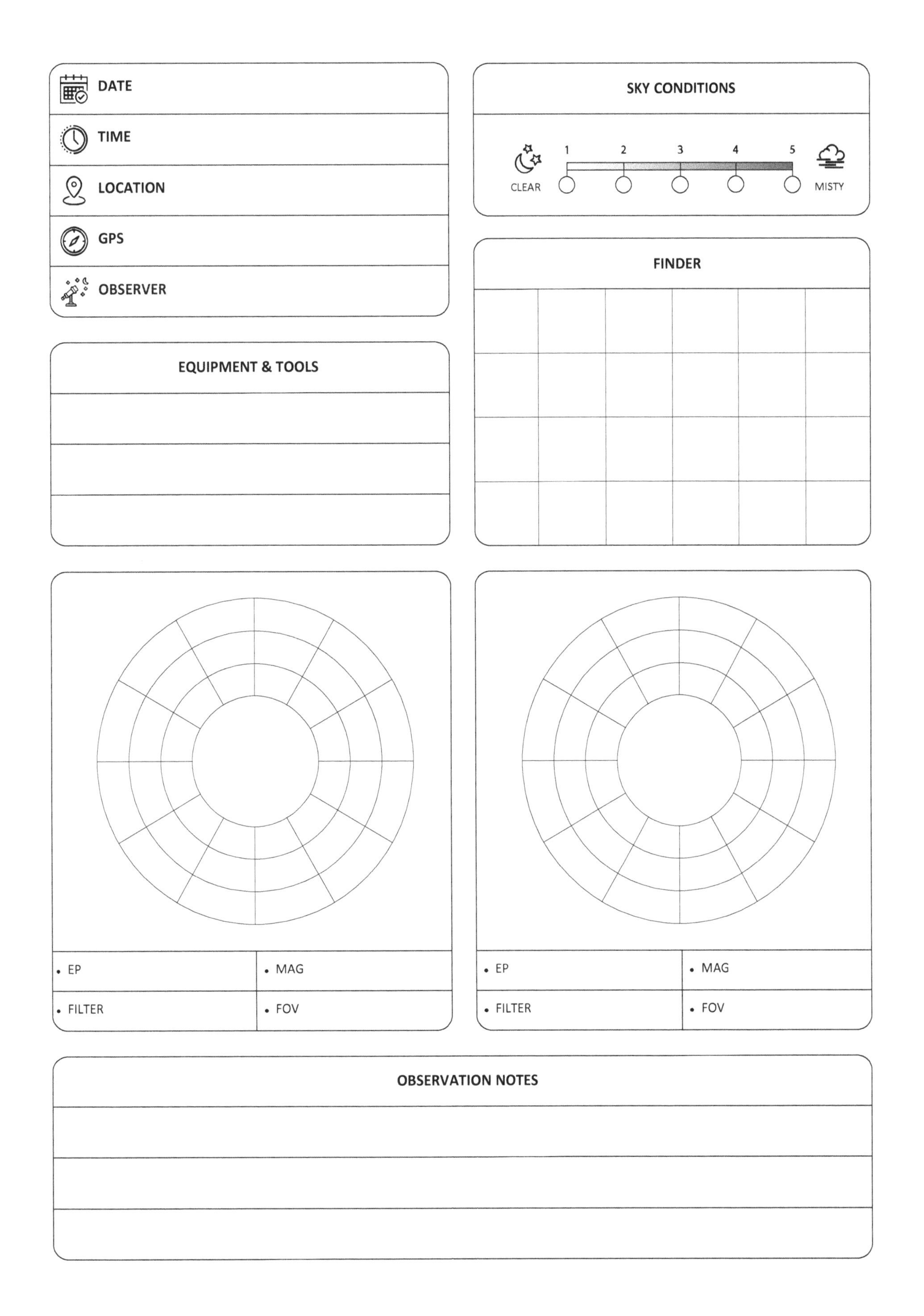
DATE
TIME
LOCATION
GPS
OBSERVER
SKY CONDITIONS
1
2
3
4
5
CLEAR
MISTY
FINDER
EQUIPMENT & TOOLS
EP
MAG
FILTER
FOV
EP
MAG
FILTER
FOV
OBSERVATION NOTES

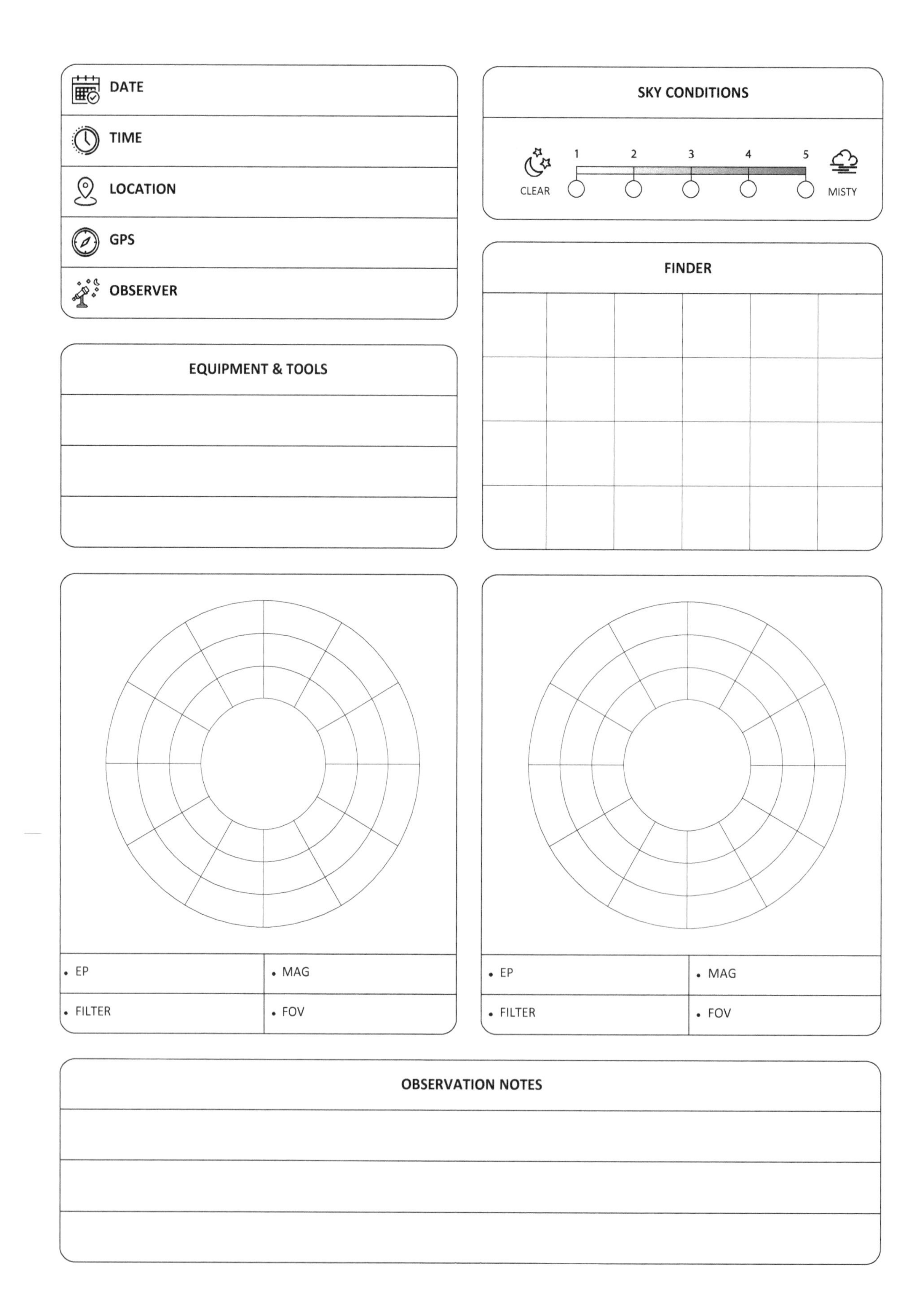

DATE
TIME
LOCATION
GPS
OBSERVER
SKY CONDITIONS
1
2
3
4
5
CLEAR
MISTY
FINDER
EQUIPMENT & TOOLS
EP
MAG
FILTER
FOV
EP
MAG
FILTER
FOV
OBSERVATION NOTES

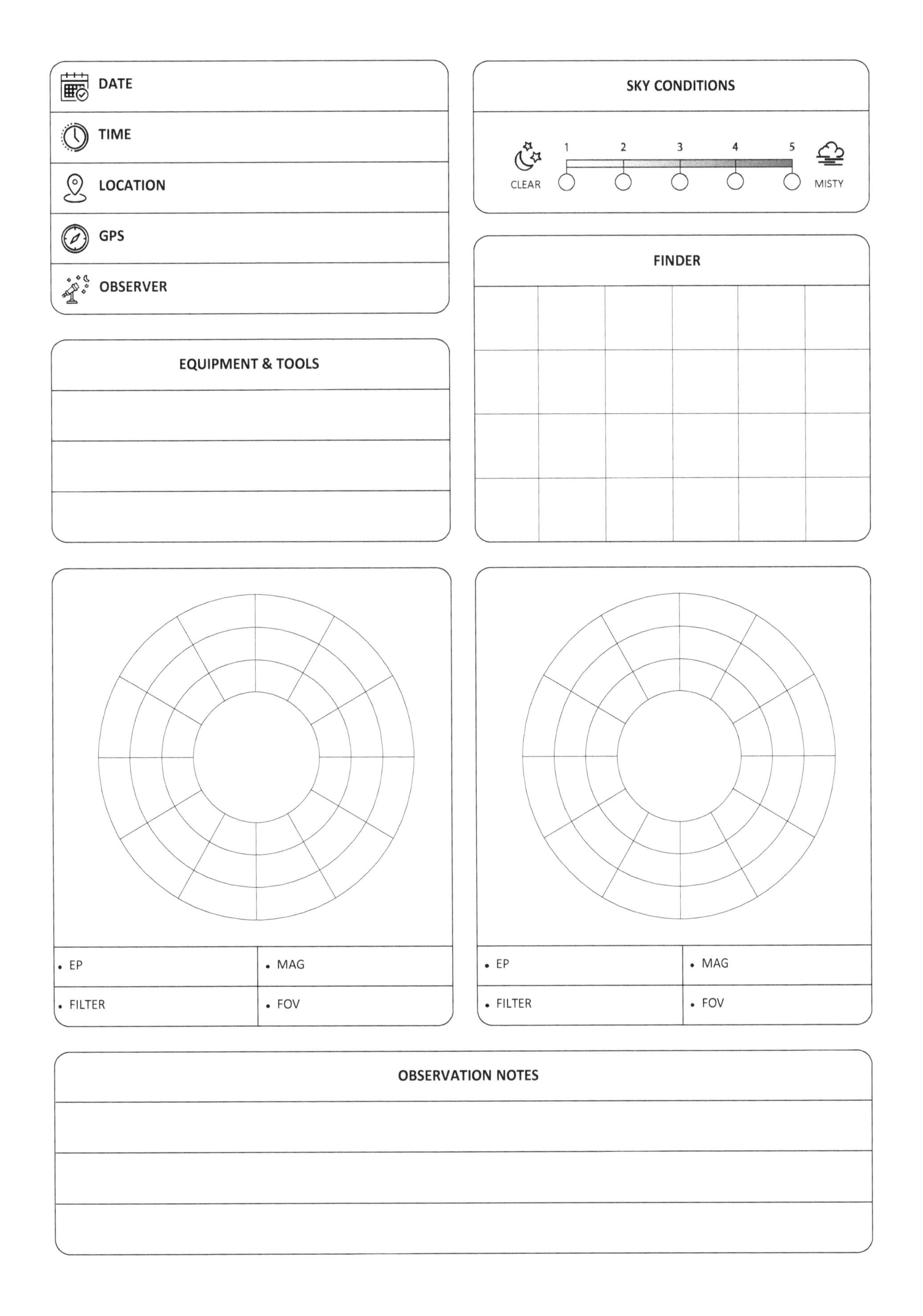
DATE
TIME
LOCATION
GPS
OBSERVER
SKY CONDITIONS
1
2
3
4
5
CLEAR
MISTY
FINDER
EQUIPMENT & TOOLS
EP
MAG
FILTER
FOV
EP
MAG
FILTER
FOV
OBSERVATION NOTES

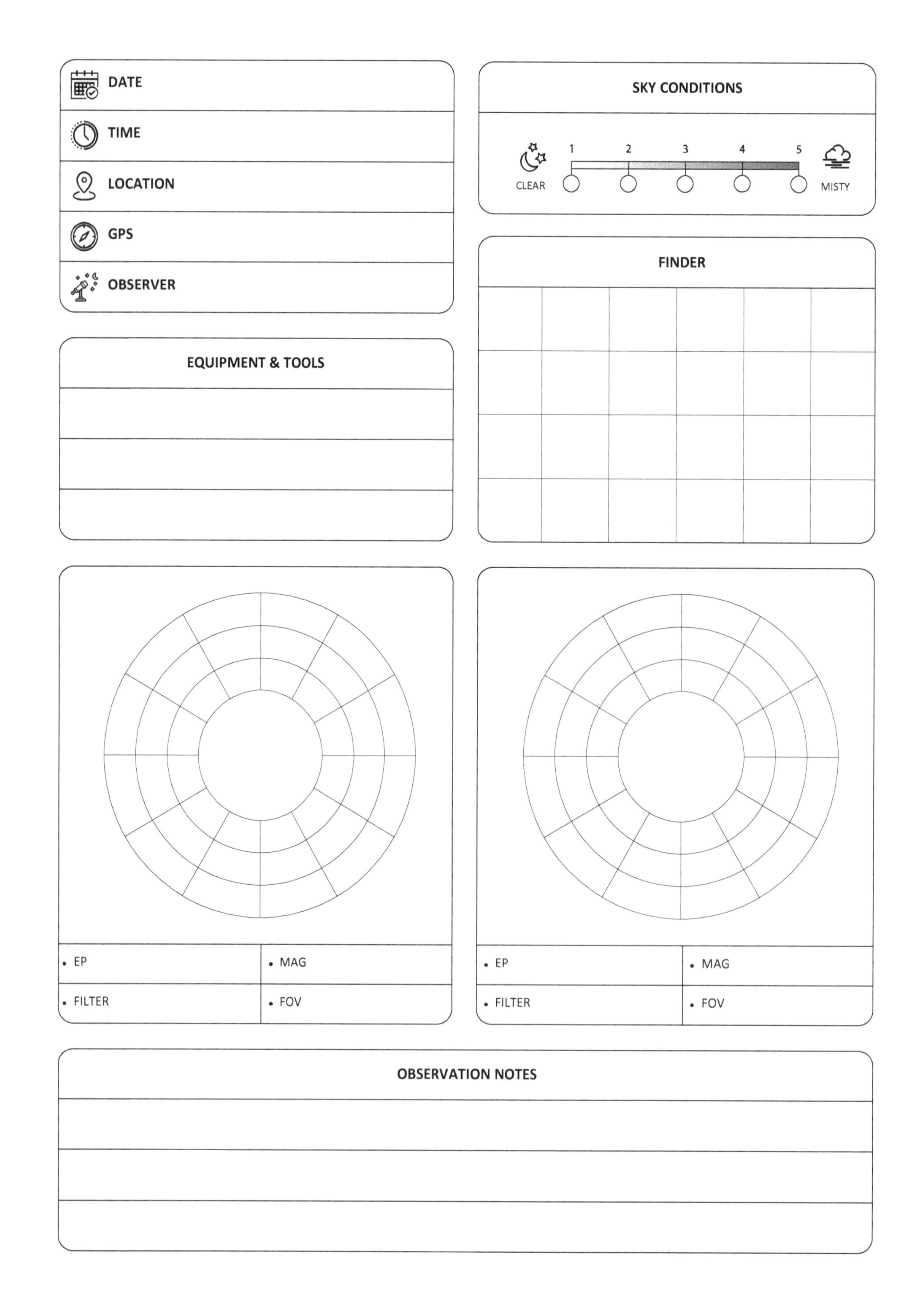

DATE
TIME
LOCATION
GPS
OBSERVER
SKY CONDITIONS
1
2
3
4
5
CLEAR
MISTY
FINDER
EQUIPMENT & TOOLS
EP
MAG
FILTER
FOV
EP
MAG
FILTER
FOV
OBSERVATION NOTES

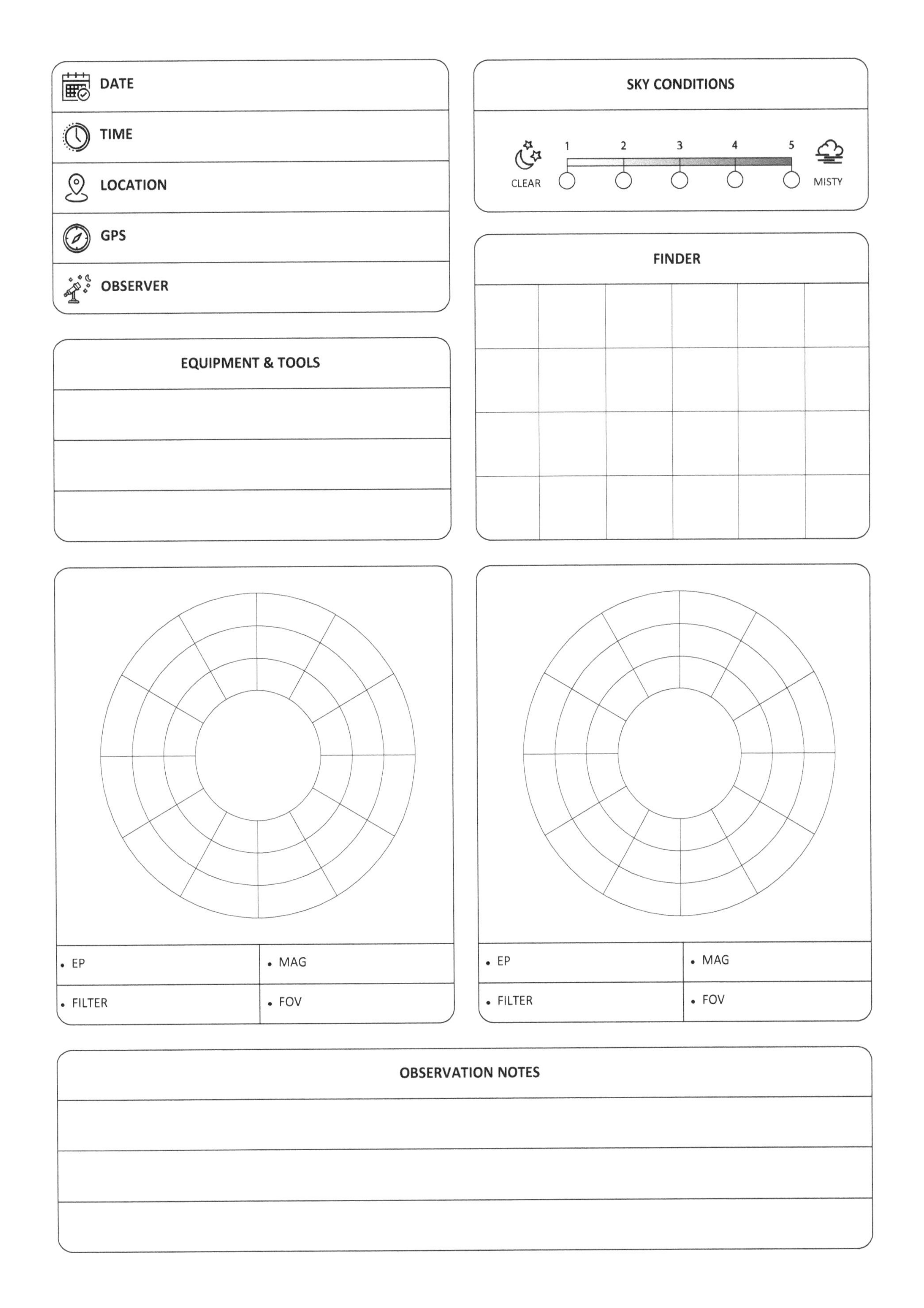
DATE
TIME
LOCATION
GPS
OBSERVER
SKY CONDITIONS
1
2
3
4
5
CLEAR
MISTY
FINDER
EQUIPMENT & TOOLS
EP
MAG
FILTER
FOV
EP
MAG
FILTER
FOV
OBSERVATION NOTES

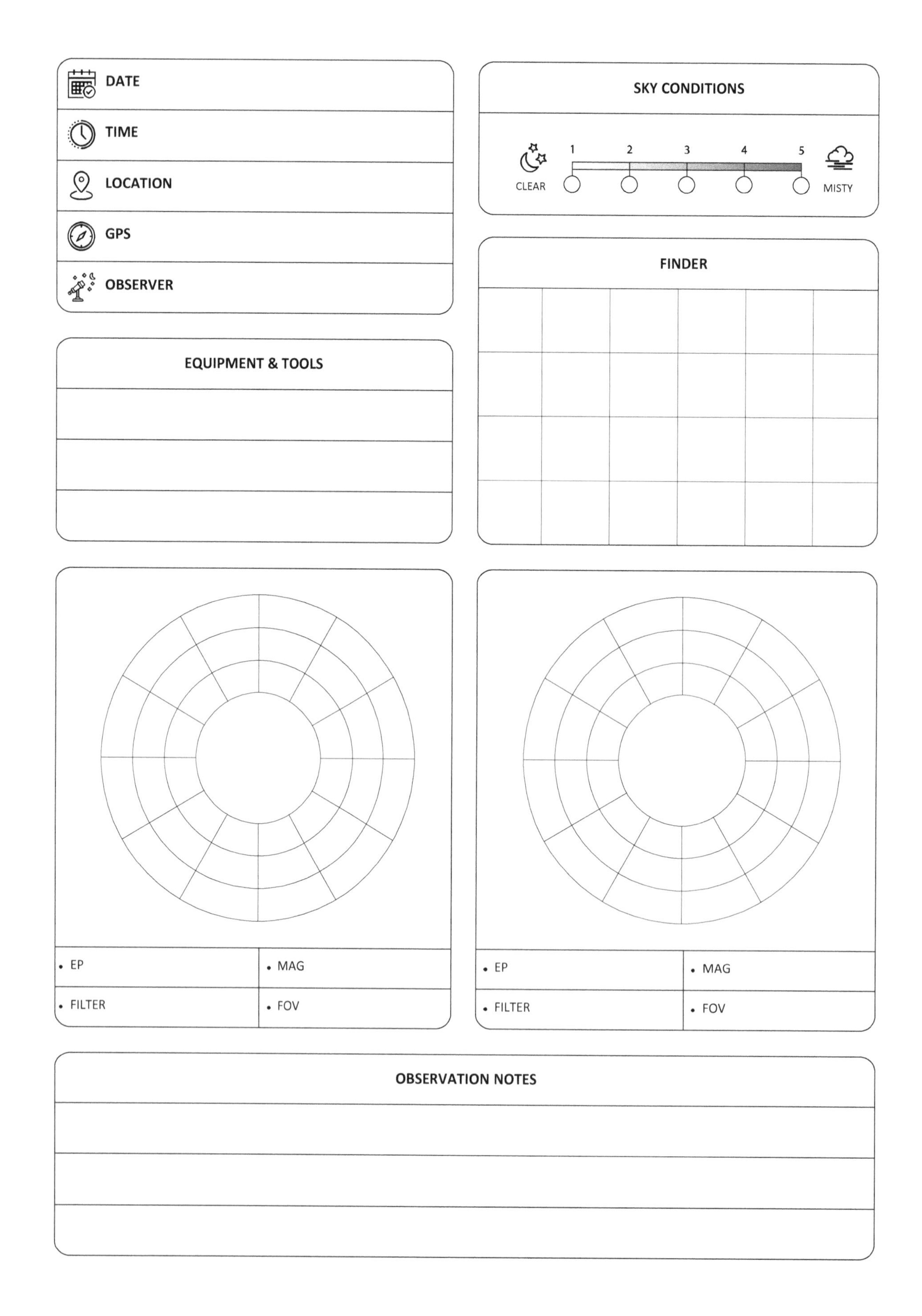
DATE
TIME
LOCATION
GPS
OBSERVER
SKY CONDITIONS
CLEAR
1
2
3
4
5
MISTY
FINDER
EQUIPMENT & TOOLS
• EP
• MAG
• FILTER
• FOV
• EP
• MAG
• FILTER
• FOV
OBSERVATION NOTES

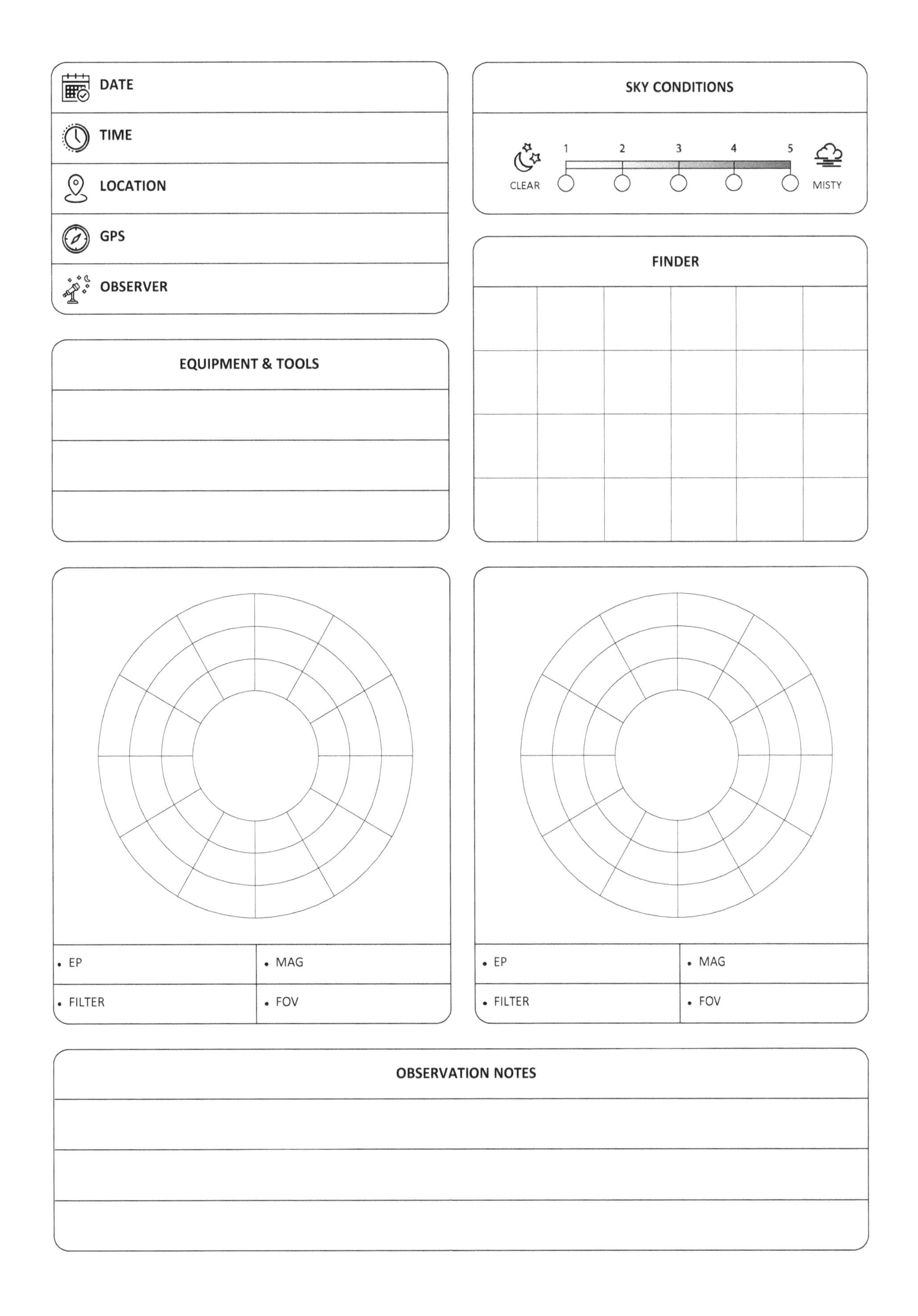
DATE
TIME
LOCATION
GPS
OBSERVER
SKY CONDITIONS
1
2
3
4
5
CLEAR
MISTY
FINDER
EQUIPMENT & TOOLS
EP
MAG
FILTER
FOV
EP
MAG
FILTER
FOV
OBSERVATION NOTES

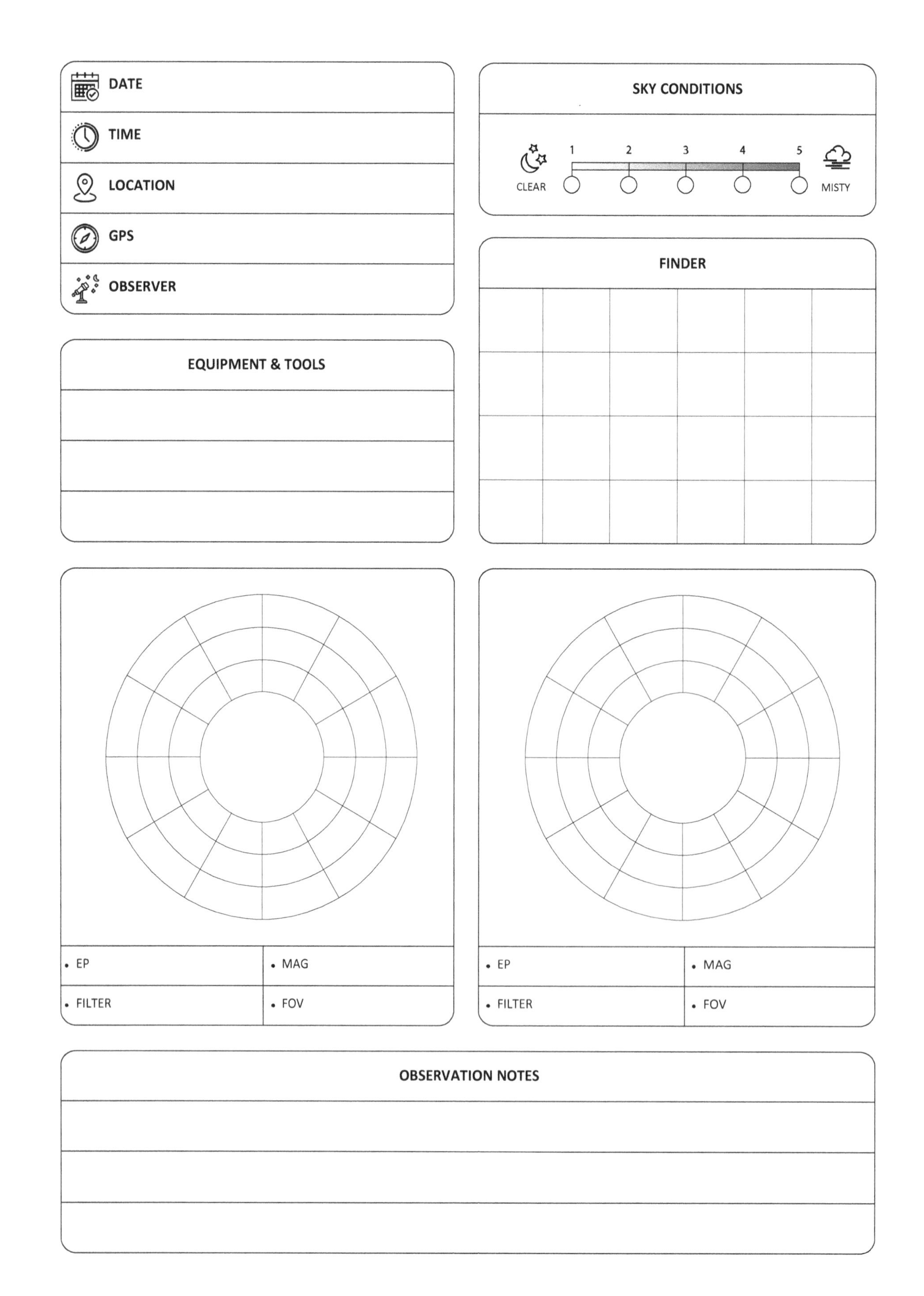
DATE
TIME
LOCATION
GPS
OBSERVER
SKY CONDITIONS
1
2
3
4
5
CLEAR
MISTY
FINDER
EQUIPMENT & TOOLS
EP
MAG
FILTER
FOV
EP
MAG
FILTER
FOV
OBSERVATION NOTES

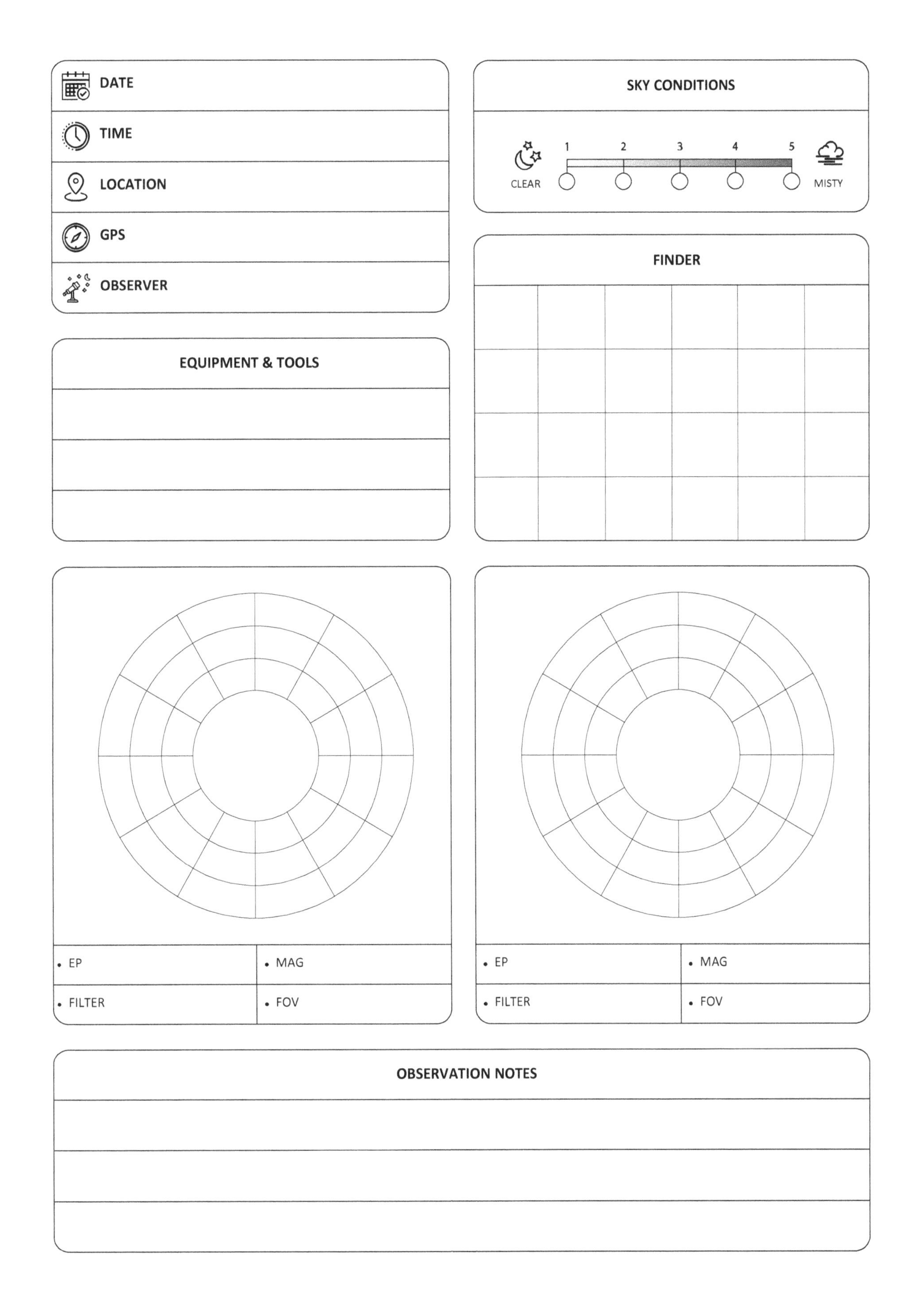

DATE

TIME

LOCATION

GPS

OBSERVER

EQUIPMENT & TOOLS

SKY CONDITIONS

CLEAR 1 2 3 4 5 MISTY

FINDER

• EP	• MAG
• FILTER	• FOV

• EP	• MAG
• FILTER	• FOV

OBSERVATION NOTES

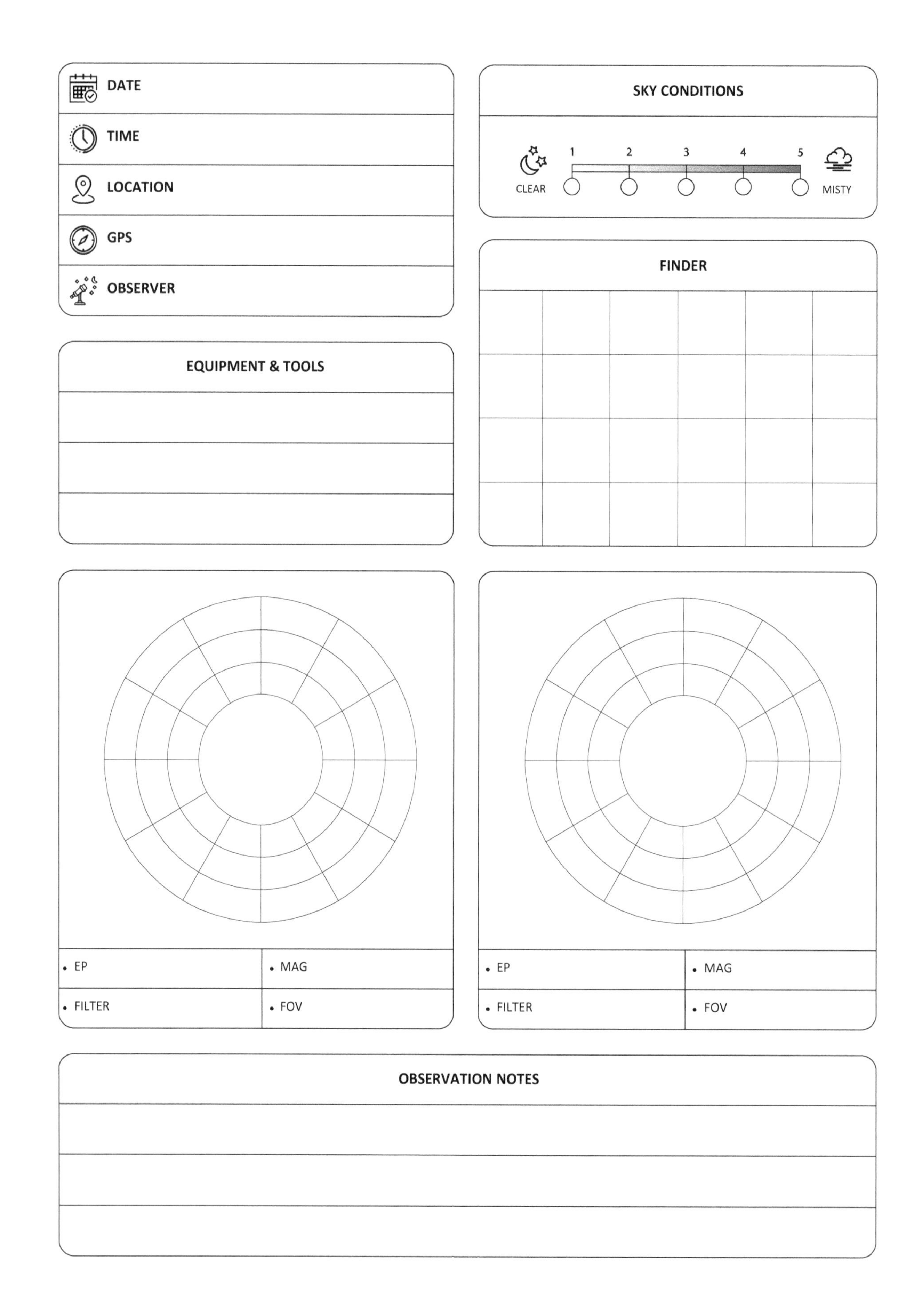
DATE
TIME
LOCATION
GPS
OBSERVER
SKY CONDITIONS
1
2
3
4
5
CLEAR
MISTY
FINDER
EQUIPMENT & TOOLS
• EP
• MAG
• FILTER
• FOV
• EP
• MAG
• FILTER
• FOV
OBSERVATION NOTES

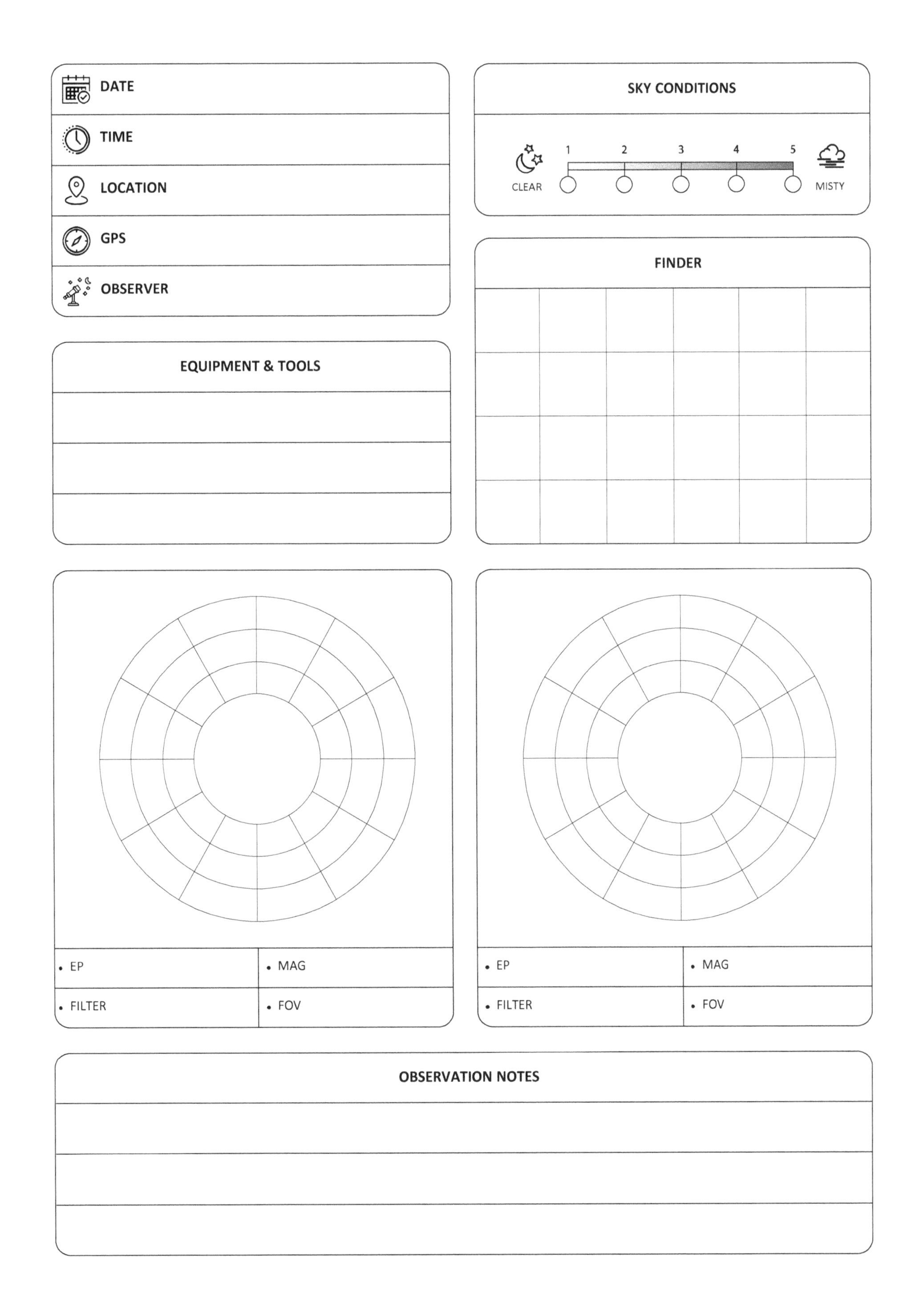
DATE
TIME
LOCATION
GPS
OBSERVER
SKY CONDITIONS
1
2
3
4
5
CLEAR
MISTY
EQUIPMENT & TOOLS
FINDER
EP
MAG
FILTER
FOV
EP
MAG
FILTER
FOV
OBSERVATION NOTES

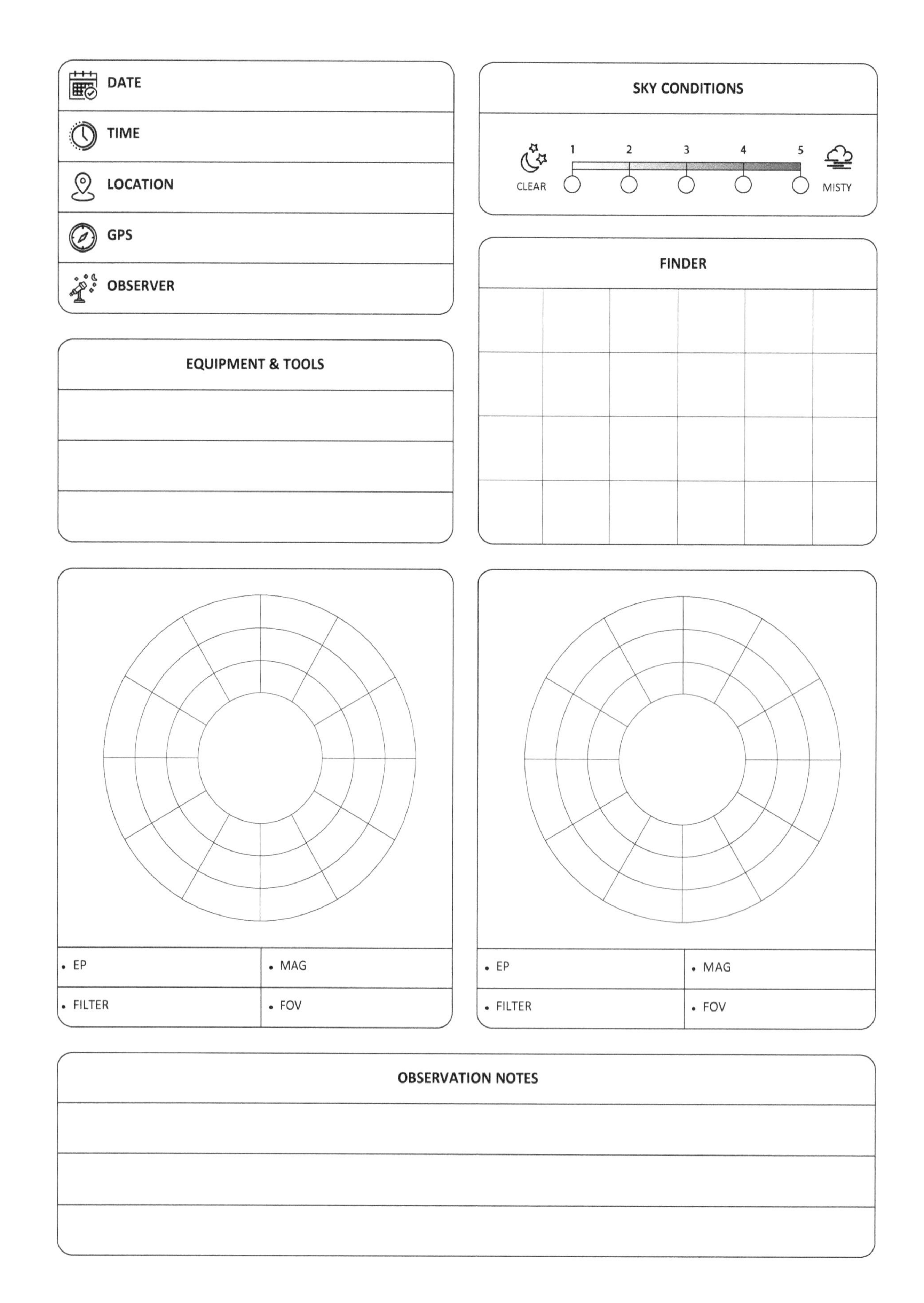
DATE
TIME
LOCATION
GPS
OBSERVER
SKY CONDITIONS
1
2
3
4
5
CLEAR
MISTY
FINDER
EQUIPMENT & TOOLS
EP
MAG
FILTER
FOV
EP
MAG
FILTER
FOV
OBSERVATION NOTES

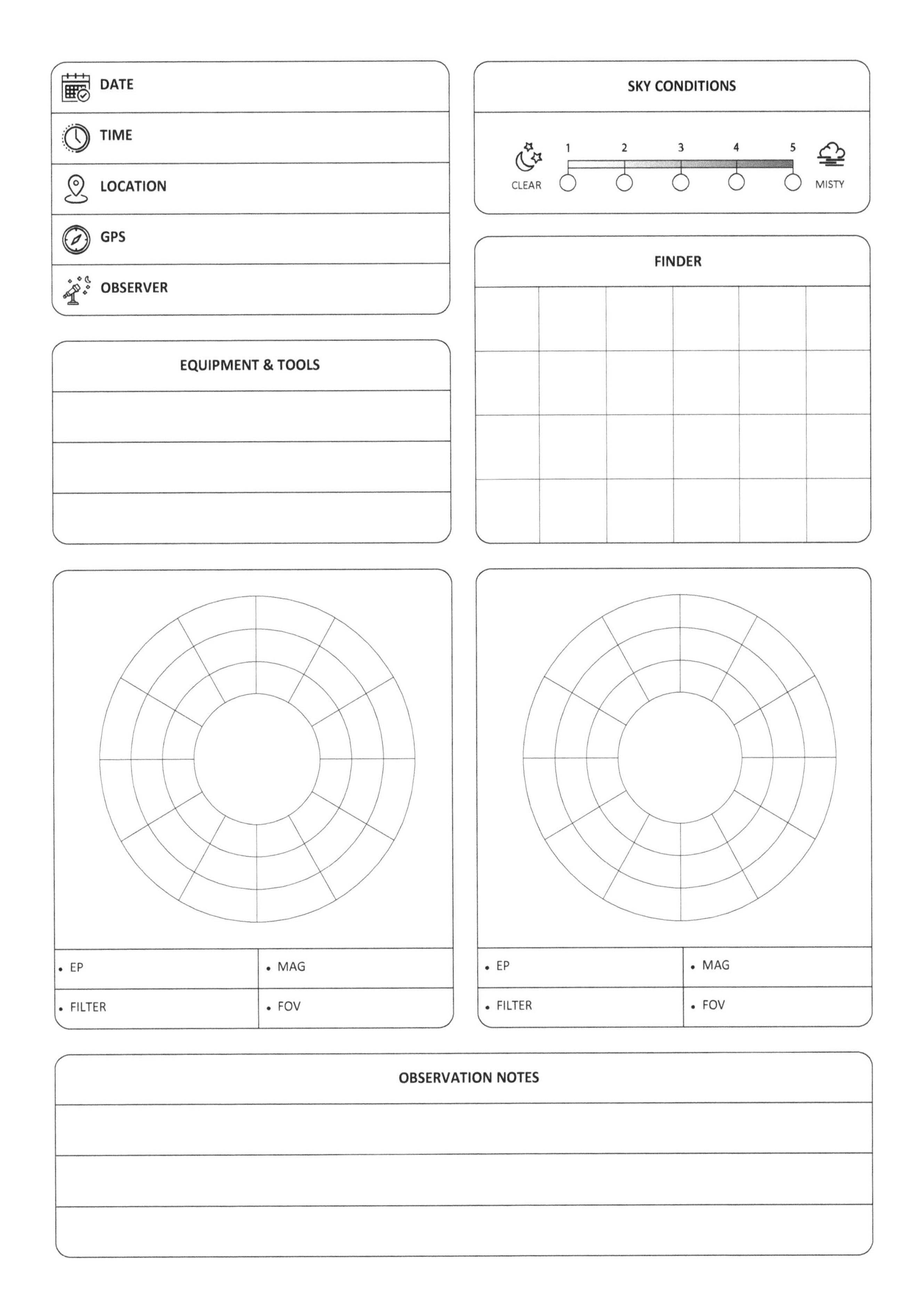

DATE

TIME

LOCATION

GPS

OBSERVER

SKY CONDITIONS

CLEAR 1 2 3 4 5 MISTY

FINDER

EQUIPMENT & TOOLS

- EP
- MAG
- FILTER
- FOV

- EP
- MAG
- FILTER
- FOV

OBSERVATION NOTES

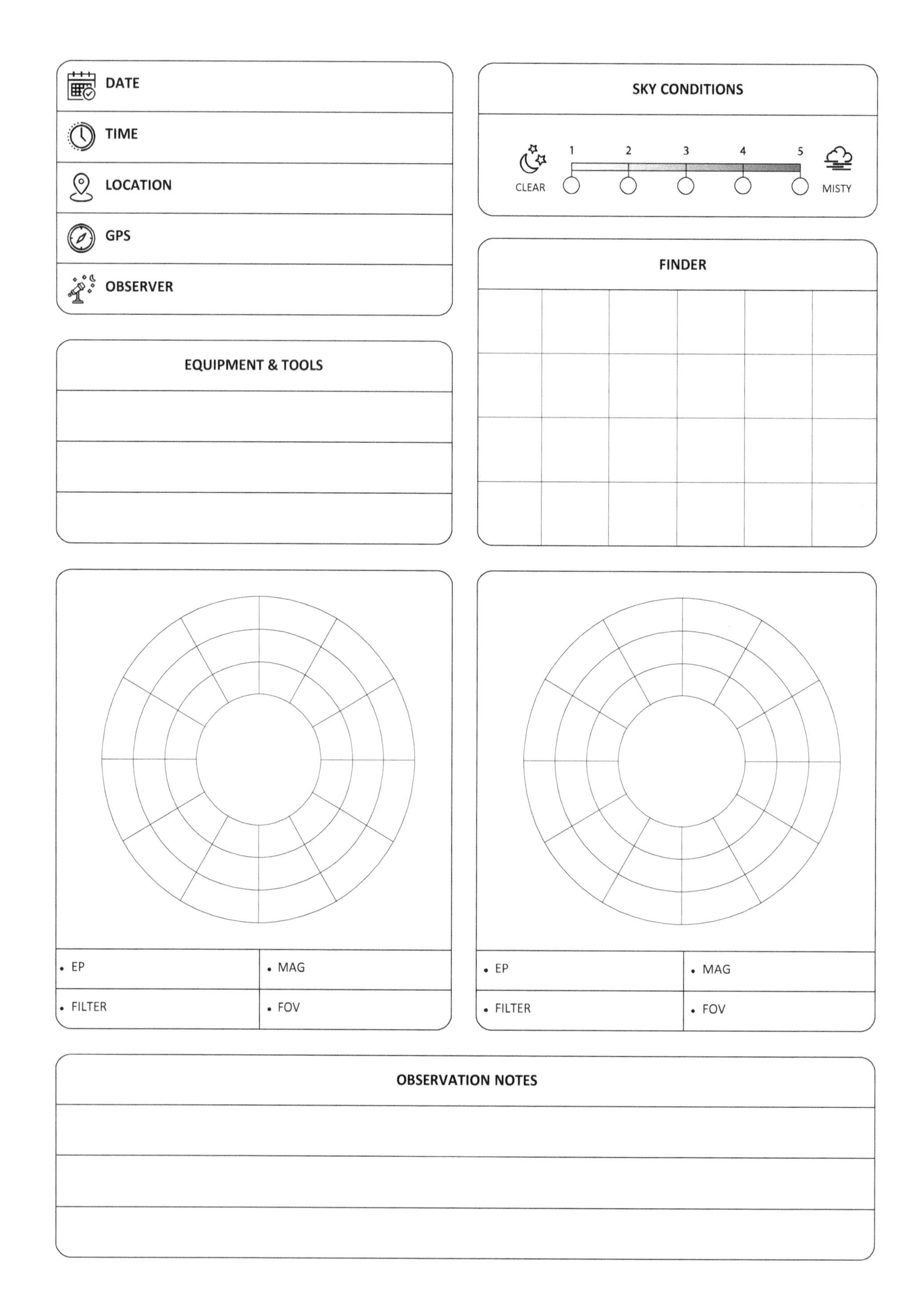
DATE
TIME
LOCATION
GPS
OBSERVER
SKY CONDITIONS
1
2
3
4
5
CLEAR
MISTY
EQUIPMENT & TOOLS
FINDER
EP
MAG
FILTER
FOV
EP
MAG
FILTER
FOV
OBSERVATION NOTES

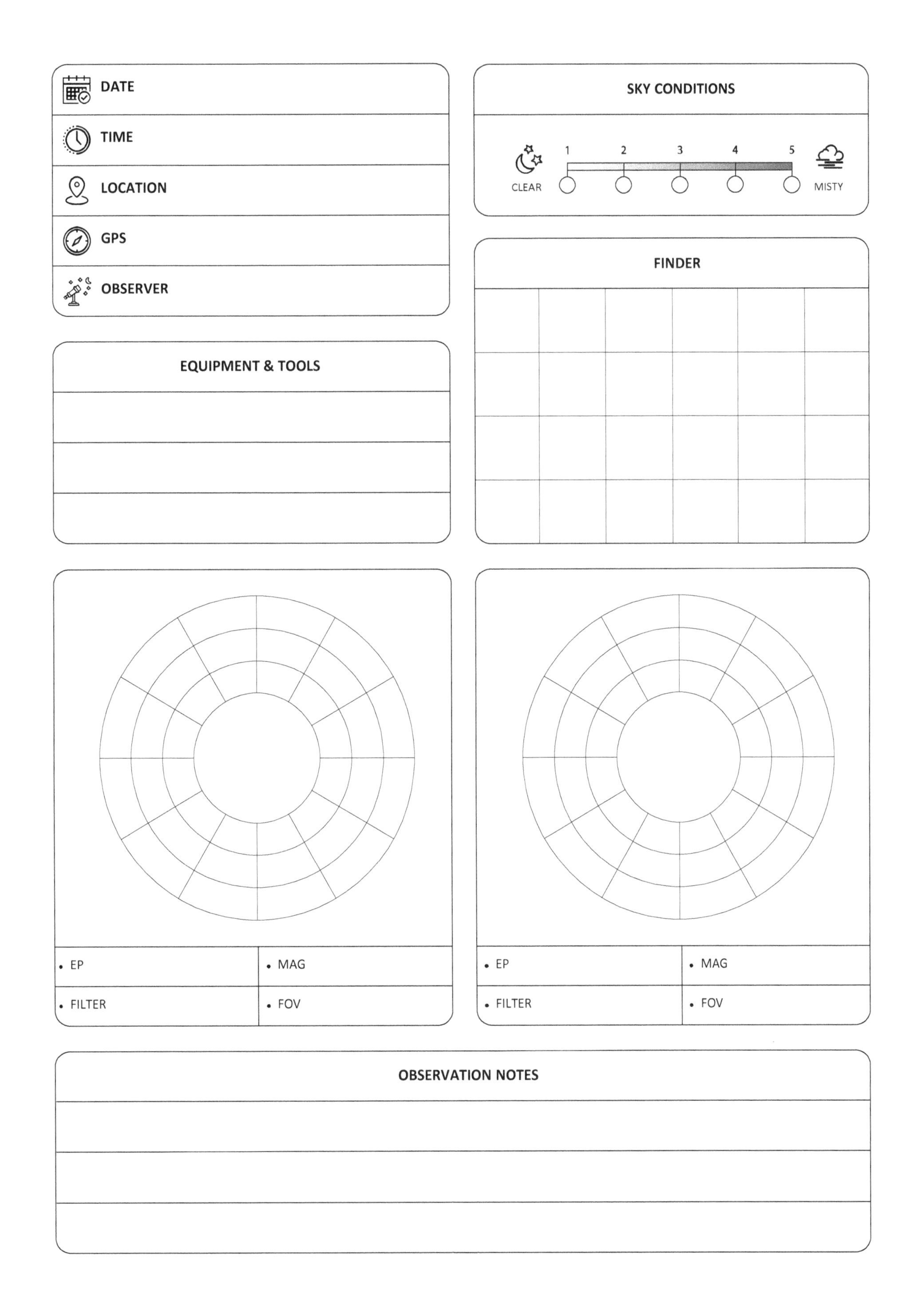
DATE
TIME
LOCATION
GPS
OBSERVER
SKY CONDITIONS
1
2
3
4
5
CLEAR
MISTY
FINDER
EQUIPMENT & TOOLS
EP
MAG
FILTER
FOV
EP
MAG
FILTER
FOV
OBSERVATION NOTES

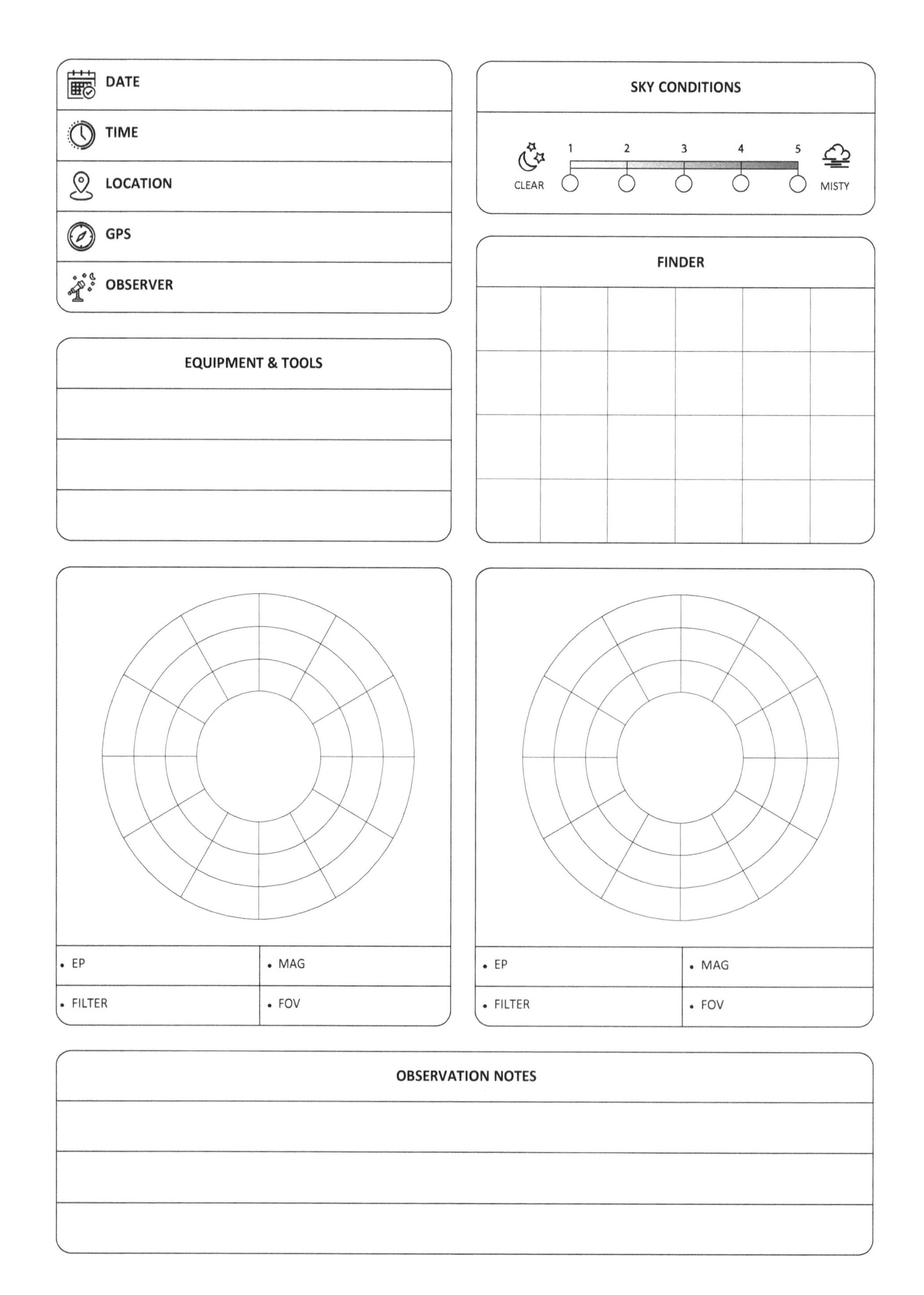
DATE
TIME
LOCATION
GPS
OBSERVER
SKY CONDITIONS
1
2
3
4
5
CLEAR
MISTY
FINDER
EQUIPMENT & TOOLS
EP
MAG
FILTER
FOV
EP
MAG
FILTER
FOV
OBSERVATION NOTES

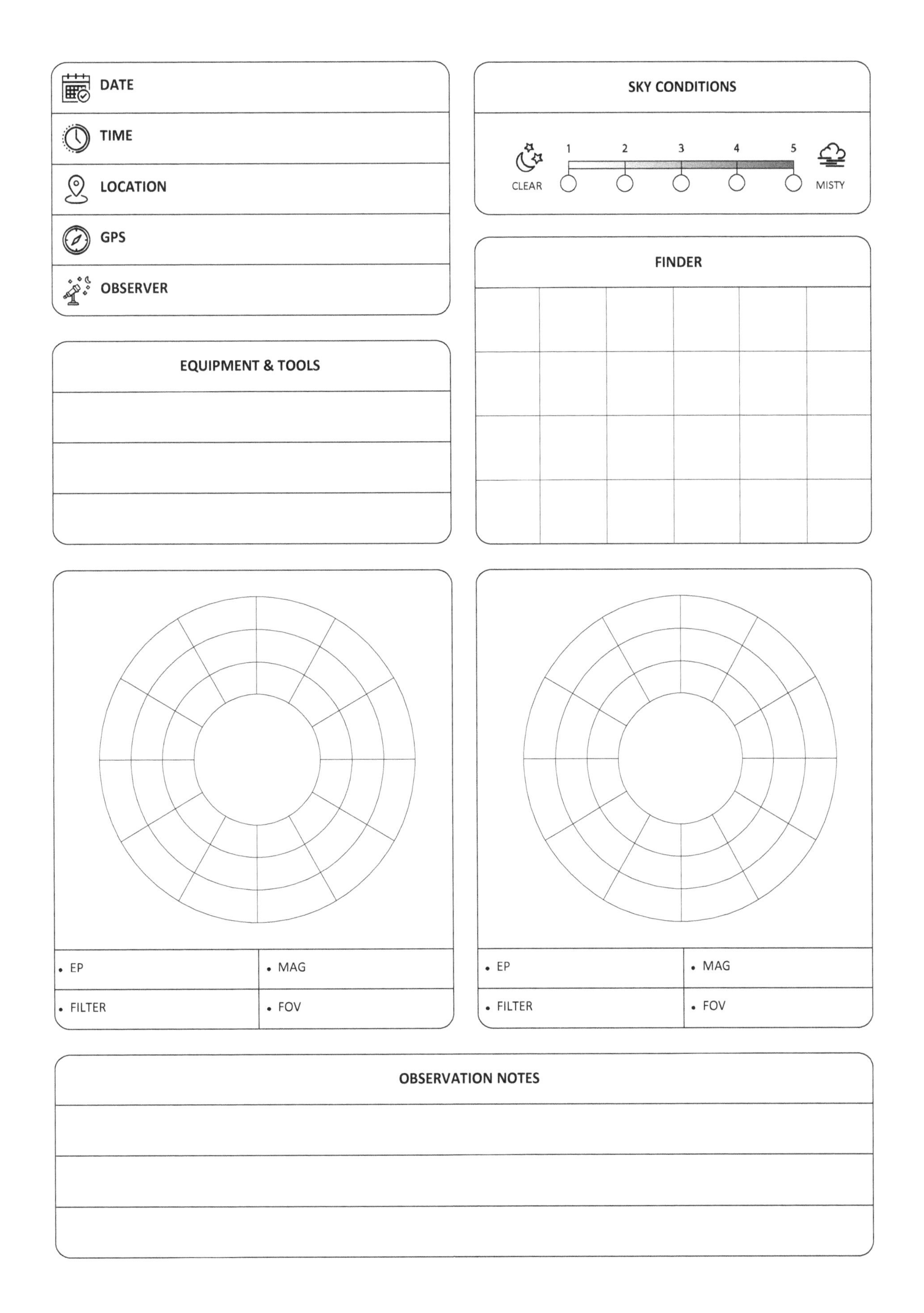
DATE
TIME
LOCATION
GPS
OBSERVER
SKY CONDITIONS
1
2
3
4
5
CLEAR
MISTY
FINDER
EQUIPMENT & TOOLS
EP
MAG
FILTER
FOV
EP
MAG
FILTER
FOV
OBSERVATION NOTES

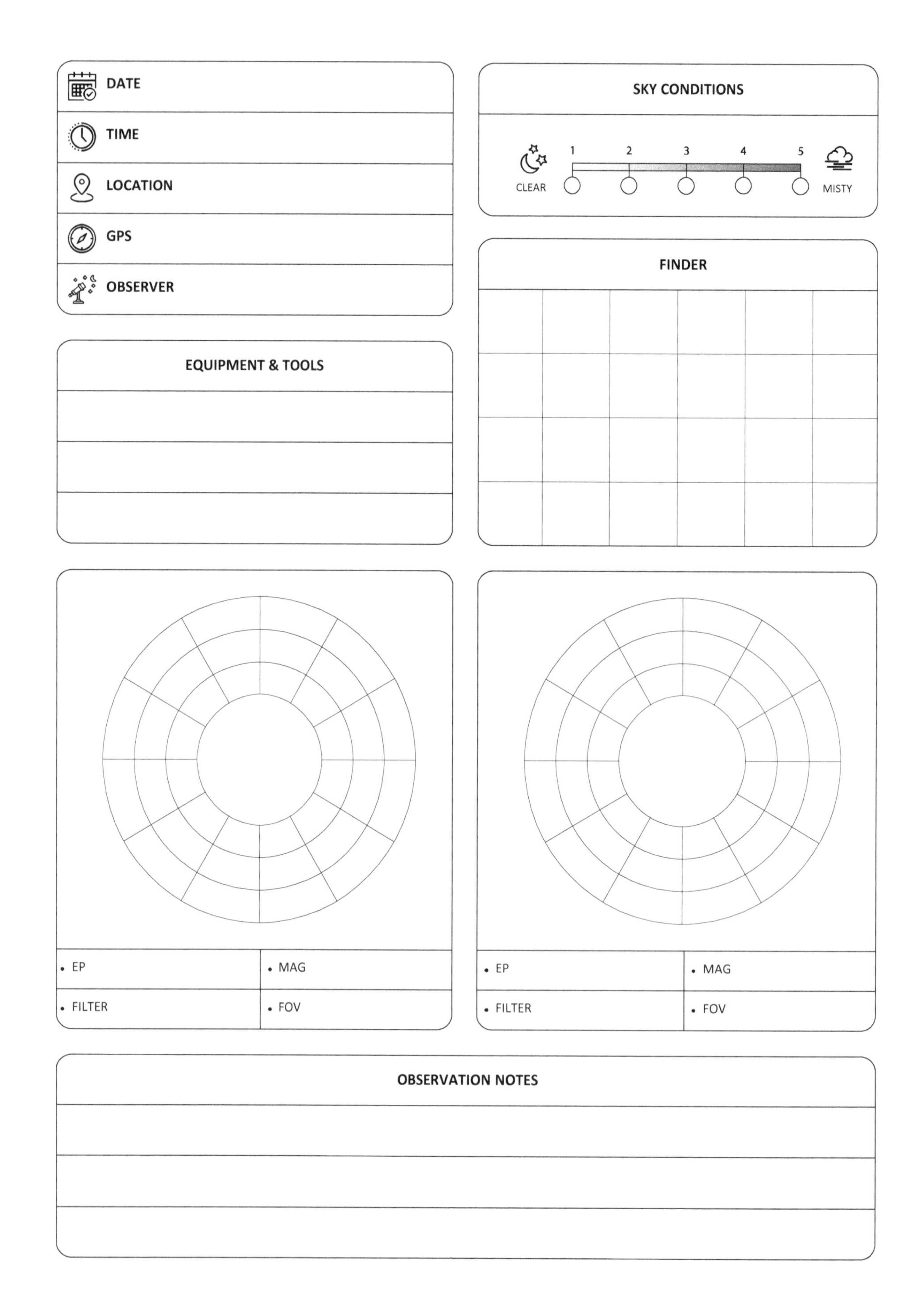
DATE
TIME
LOCATION
GPS
OBSERVER
SKY CONDITIONS
1
2
3
4
5
CLEAR
MISTY
FINDER
EQUIPMENT & TOOLS
EP
MAG
FILTER
FOV
EP
MAG
FILTER
FOV
OBSERVATION NOTES

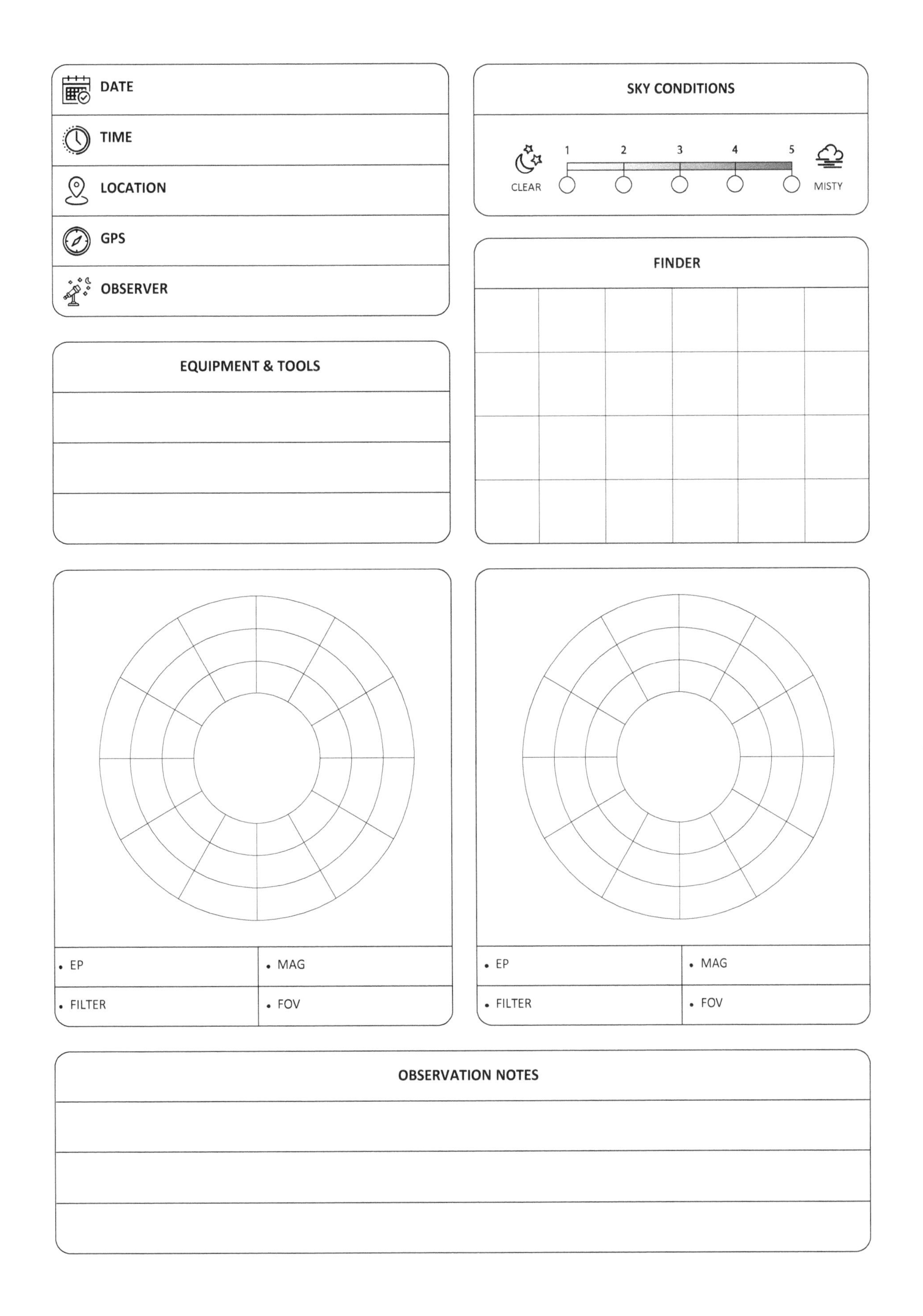
DATE
TIME
LOCATION
GPS
OBSERVER
SKY CONDITIONS
1
2
3
4
5
CLEAR
MISTY
FINDER
EQUIPMENT & TOOLS
EP
MAG
FILTER
FOV
EP
MAG
FILTER
FOV
OBSERVATION NOTES

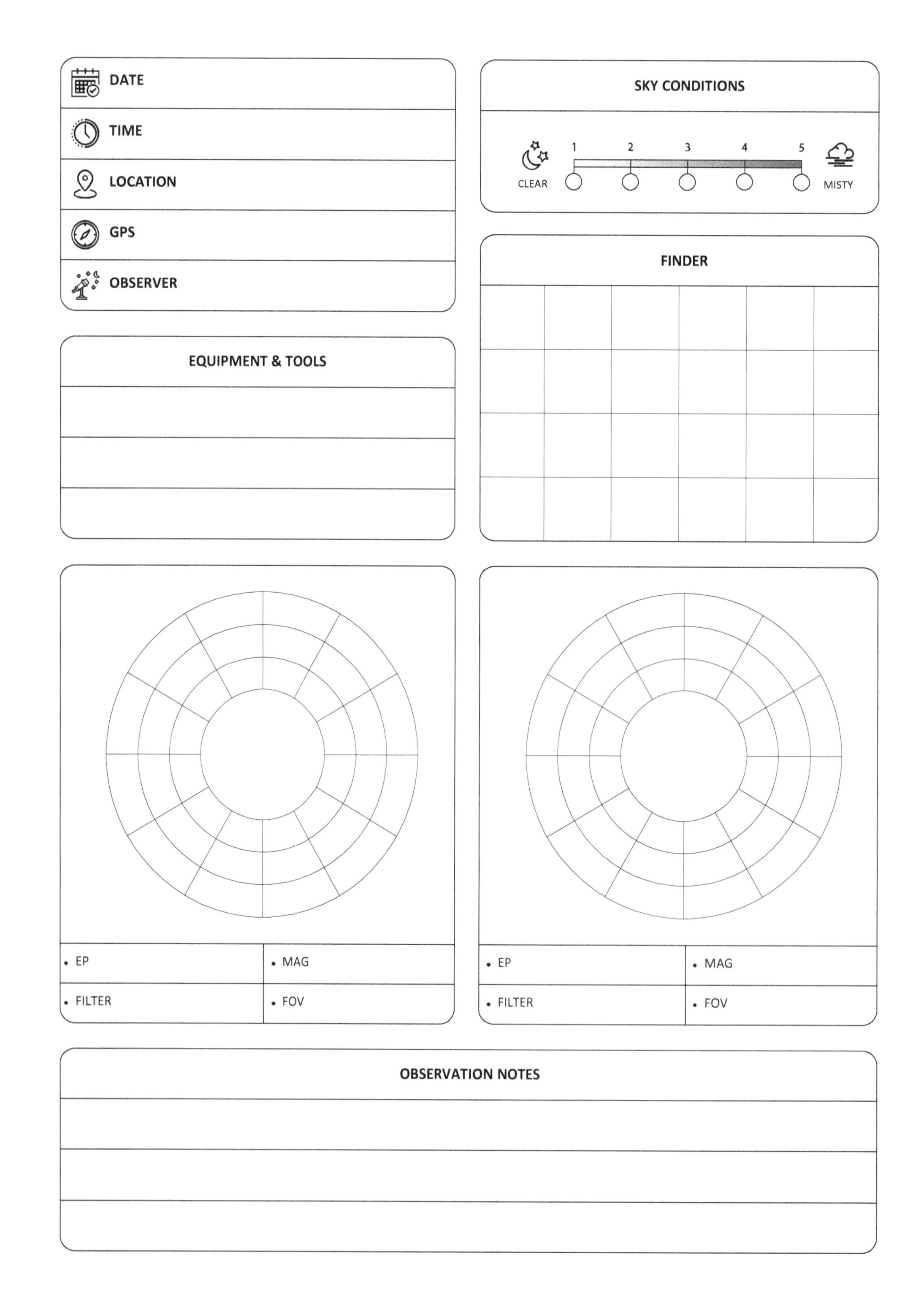
DATE
TIME
LOCATION
GPS
OBSERVER
SKY CONDITIONS
1
2
3
4
5
CLEAR
MISTY
EQUIPMENT & TOOLS
FINDER
EP
MAG
FILTER
FOV
EP
MAG
FILTER
FOV
OBSERVATION NOTES

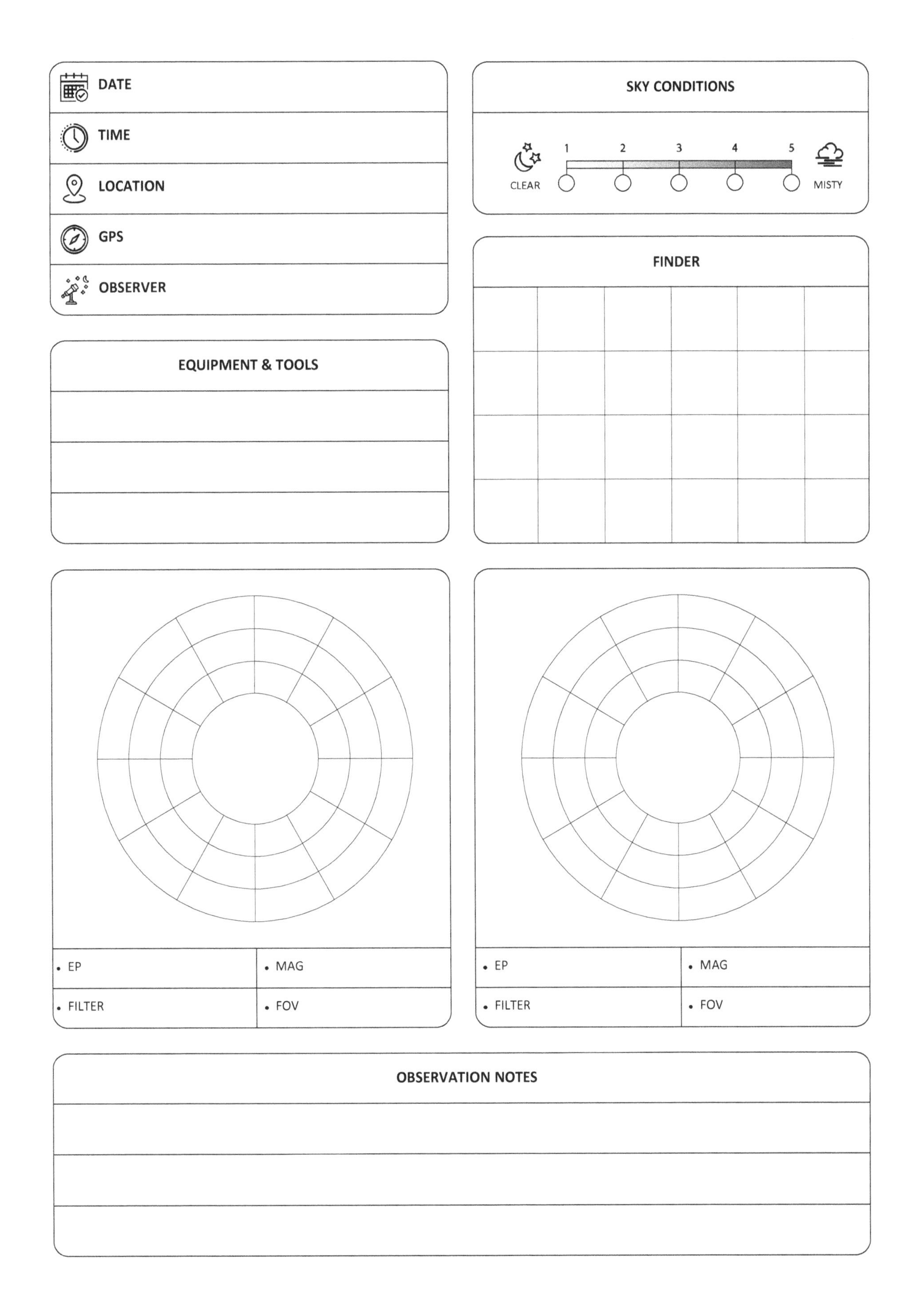
DATE
TIME
LOCATION
GPS
OBSERVER
SKY CONDITIONS
1
2
3
4
5
CLEAR
MISTY
EQUIPMENT & TOOLS
FINDER
EP
MAG
FILTER
FOV
EP
MAG
FILTER
FOV
OBSERVATION NOTES

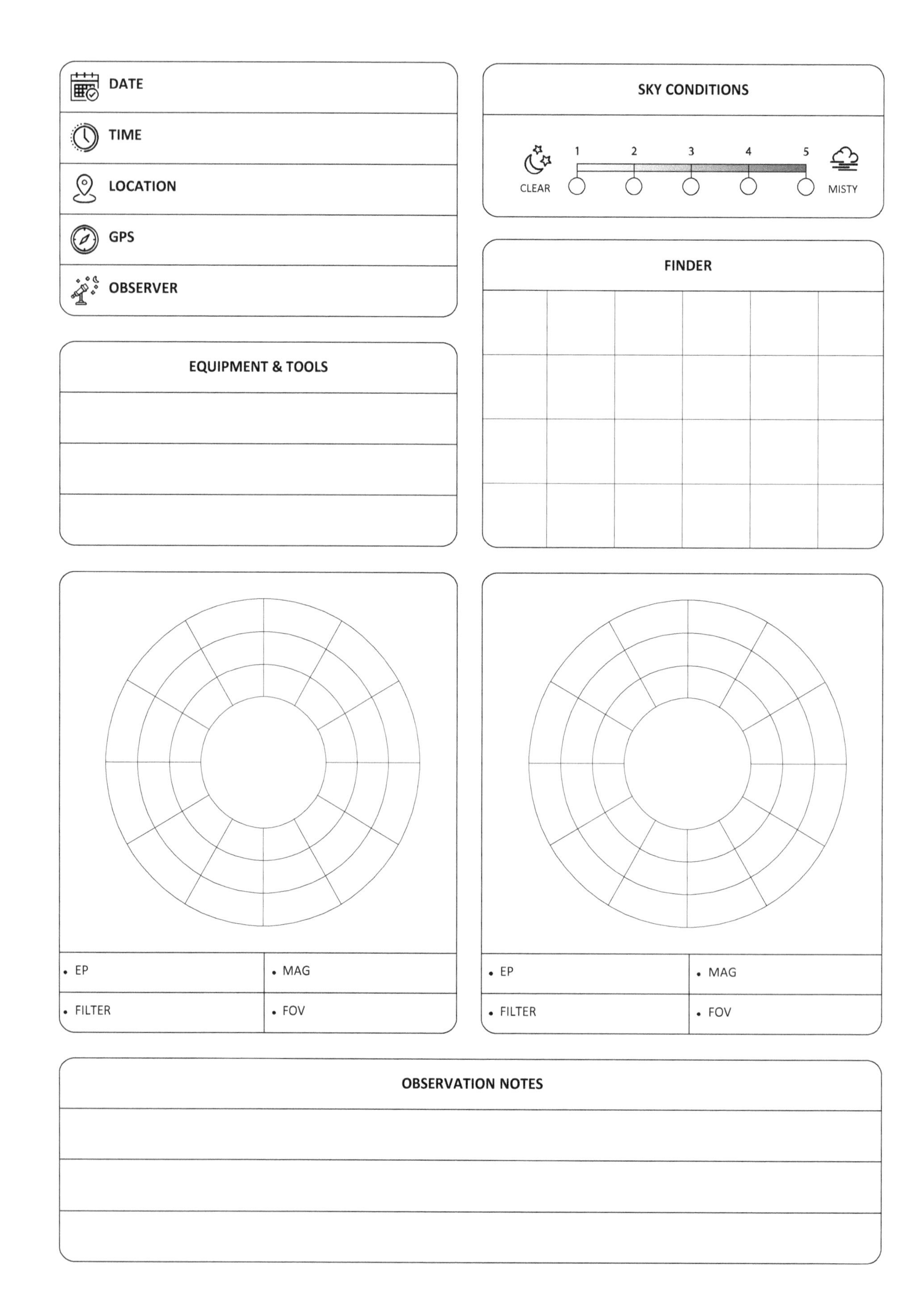
DATE
TIME
LOCATION
GPS
OBSERVER
SKY CONDITIONS
1
2
3
4
5
CLEAR
MISTY
FINDER
EQUIPMENT & TOOLS
• EP
• MAG
• FILTER
• FOV
• EP
• MAG
• FILTER
• FOV
OBSERVATION NOTES

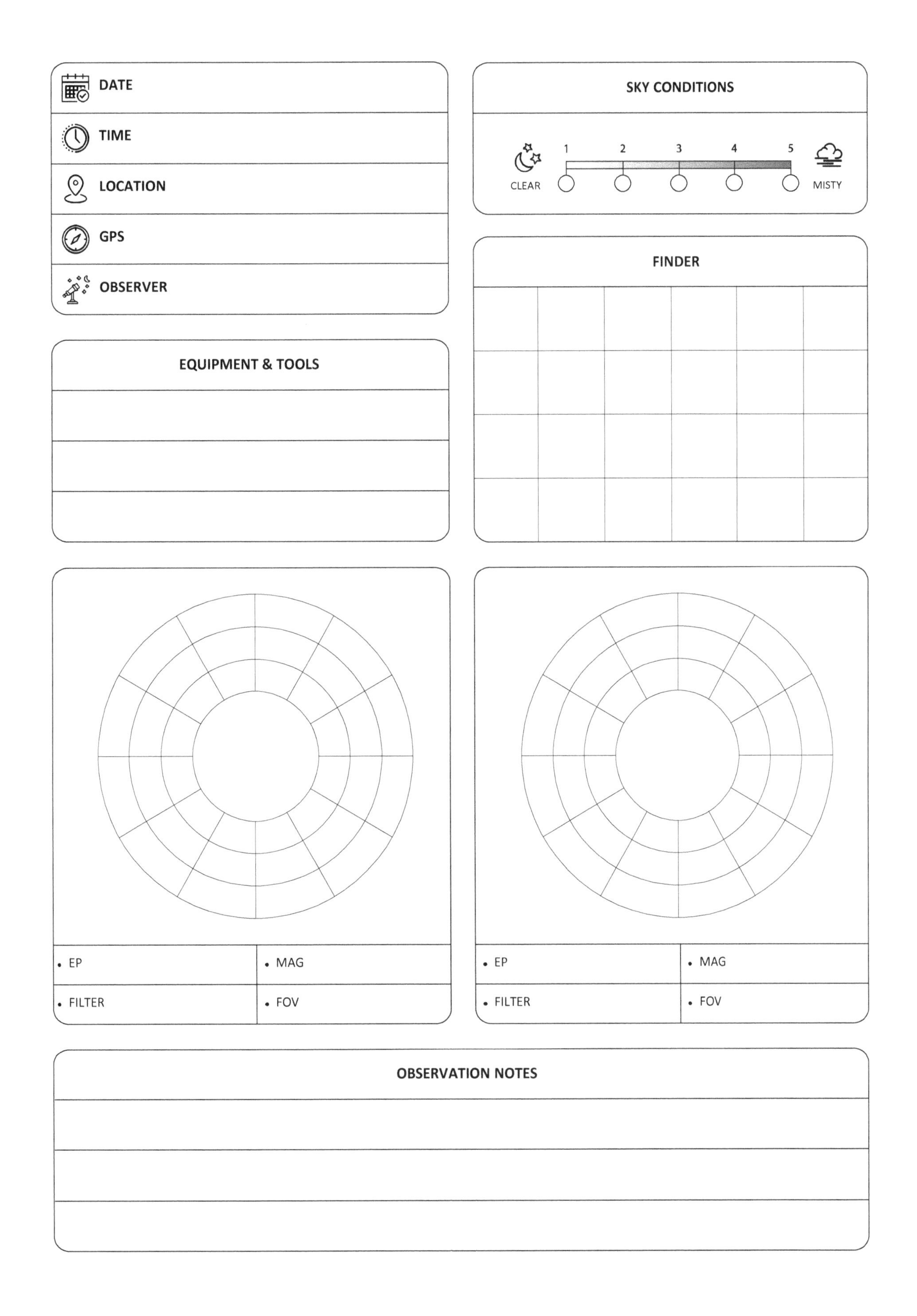
DATE
TIME
LOCATION
GPS
OBSERVER
SKY CONDITIONS
1
2
3
4
5
CLEAR
MISTY
EQUIPMENT & TOOLS
FINDER
EP
MAG
FILTER
FOV
EP
MAG
FILTER
FOV
OBSERVATION NOTES

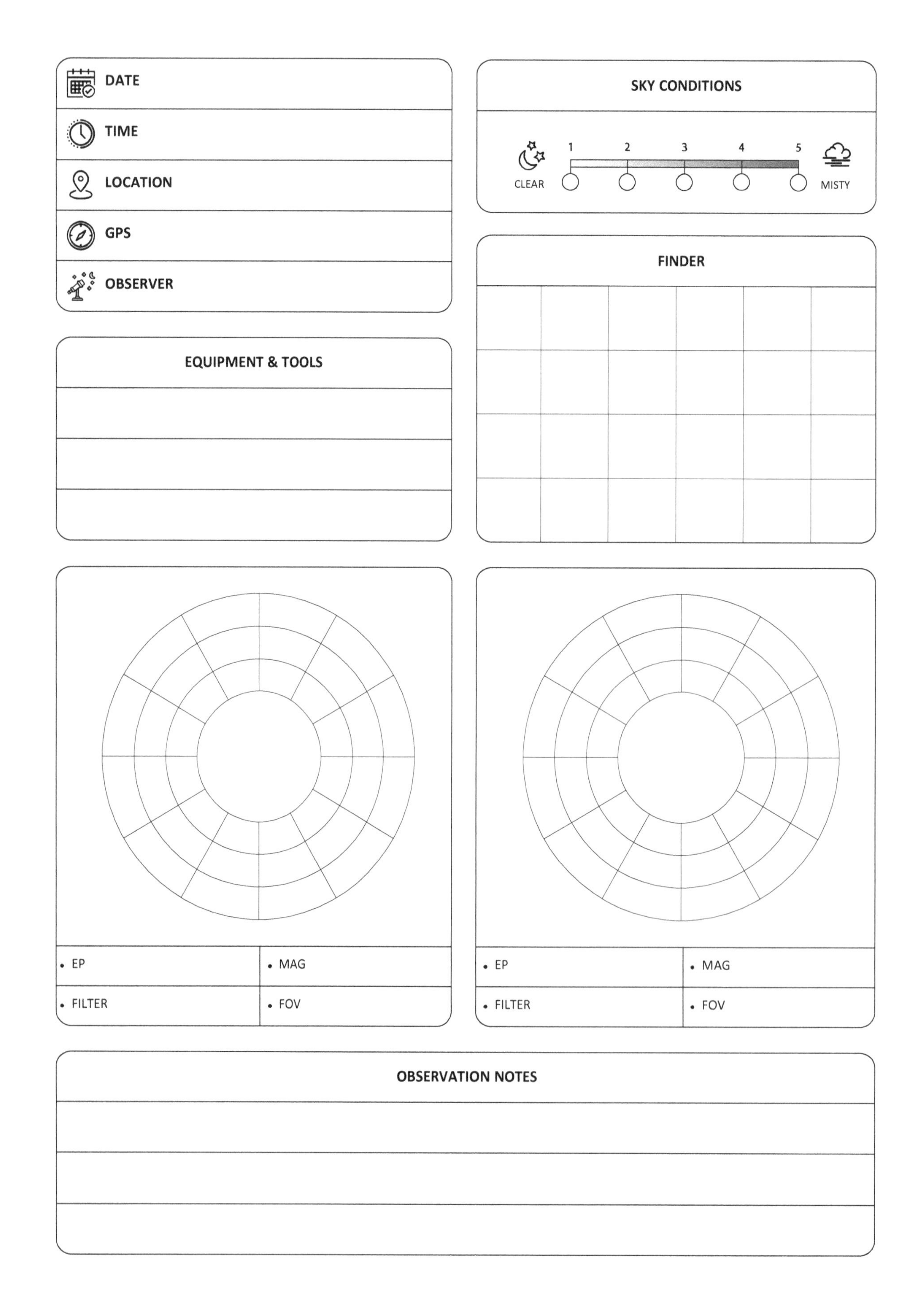
DATE
TIME
LOCATION
GPS
OBSERVER
SKY CONDITIONS
1
2
3
4
5
CLEAR
MISTY
FINDER
EQUIPMENT & TOOLS
EP
MAG
FILTER
FOV
EP
MAG
FILTER
FOV
OBSERVATION NOTES

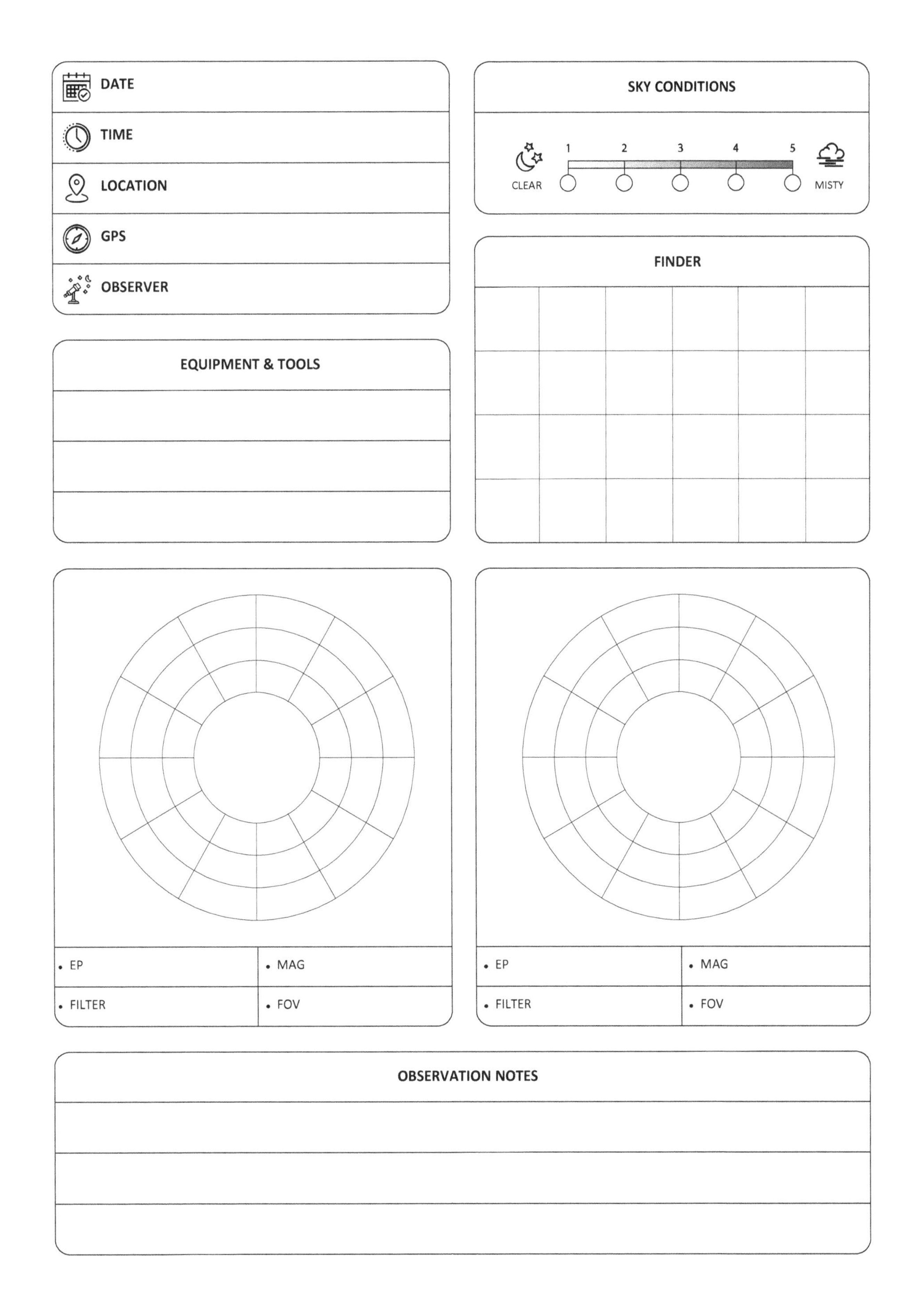
DATE
TIME
LOCATION
GPS
OBSERVER
SKY CONDITIONS
1
2
3
4
5
CLEAR
MISTY
FINDER
EQUIPMENT & TOOLS
EP
MAG
FILTER
FOV
EP
MAG
FILTER
FOV
OBSERVATION NOTES

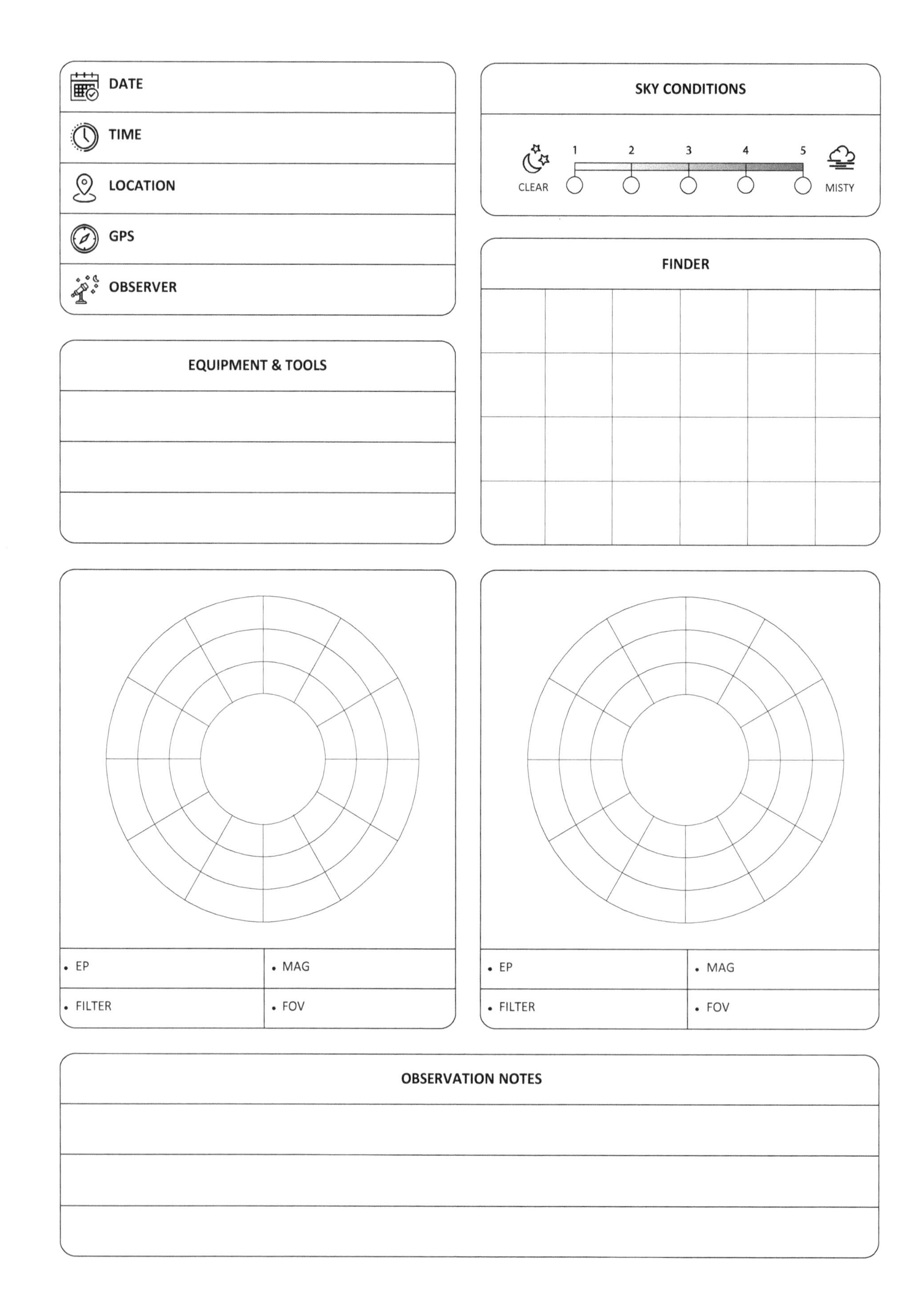

DATE
TIME
LOCATION
GPS
OBSERVER
SKY CONDITIONS
1
2
3
4
5
CLEAR
MISTY
FINDER
EQUIPMENT & TOOLS
EP
MAG
FILTER
FOV
EP
MAG
FILTER
FOV
OBSERVATION NOTES

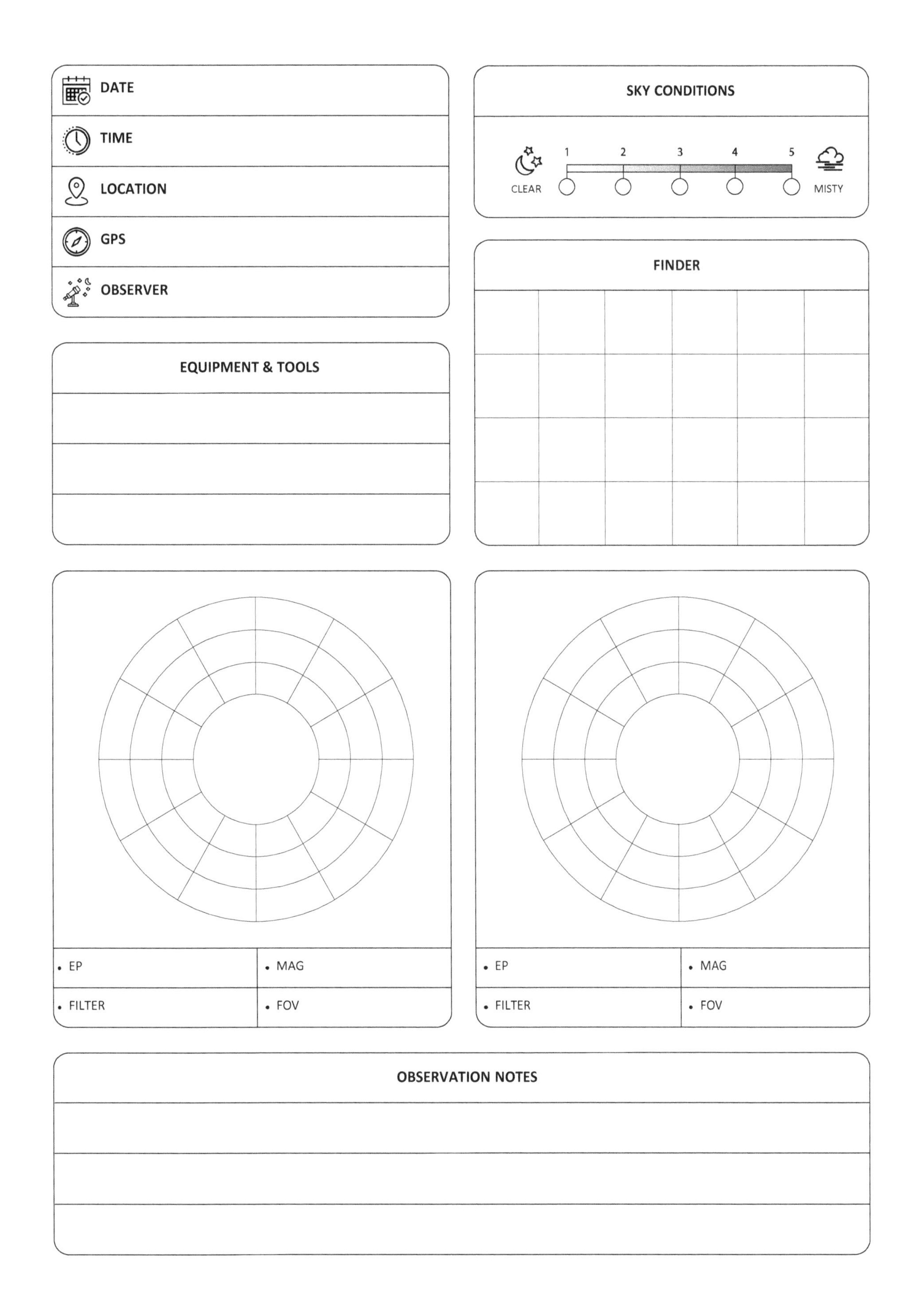
DATE
TIME
LOCATION
GPS
OBSERVER
SKY CONDITIONS
1
2
3
4
5
CLEAR
MISTY
FINDER
EQUIPMENT & TOOLS
EP
MAG
FILTER
FOV
EP
MAG
FILTER
FOV
OBSERVATION NOTES

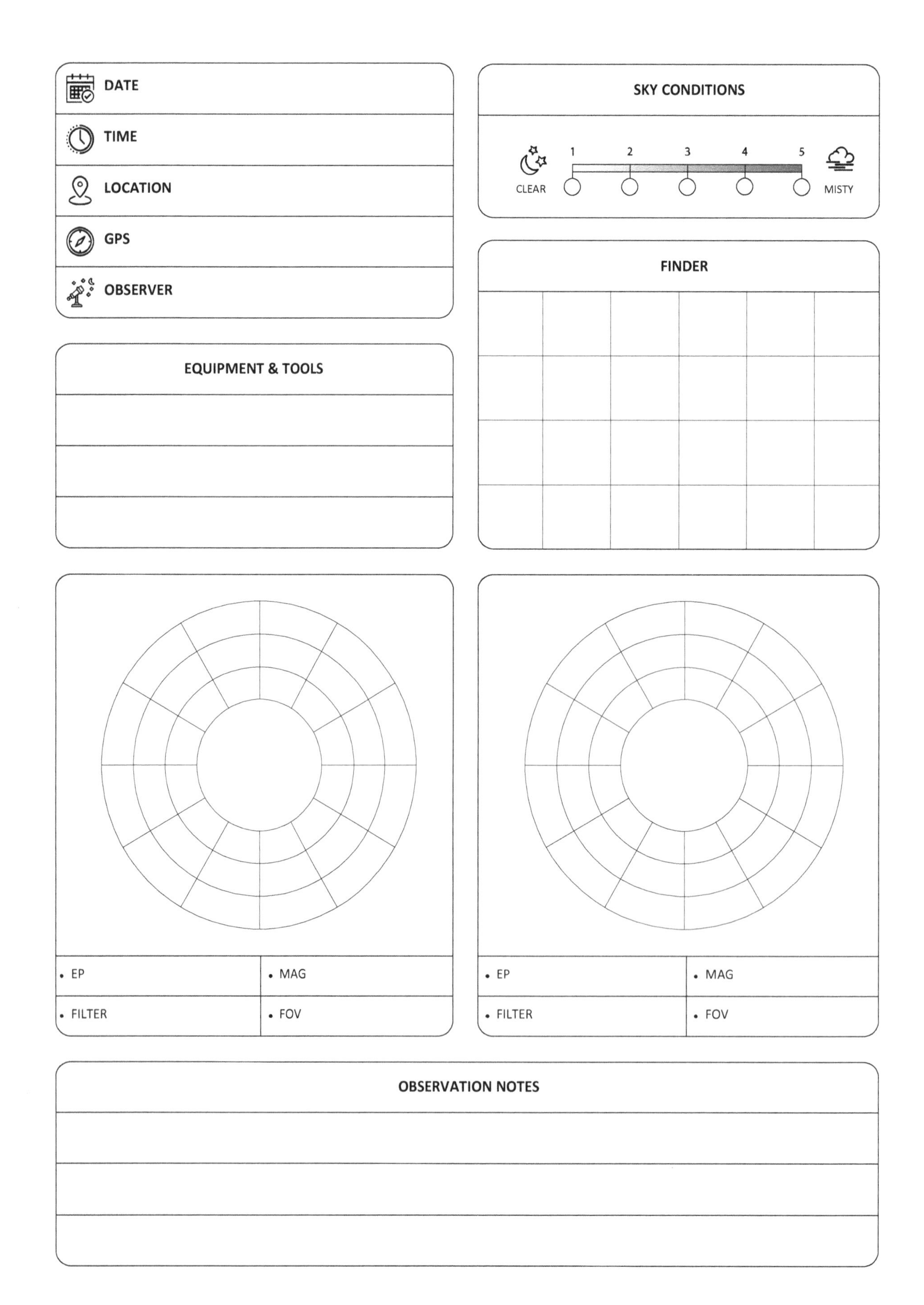
DATE
TIME
LOCATION
GPS
OBSERVER
SKY CONDITIONS
1
2
3
4
5
CLEAR
MISTY
EQUIPMENT & TOOLS
FINDER
EP
MAG
FILTER
FOV
EP
MAG
FILTER
FOV
OBSERVATION NOTES

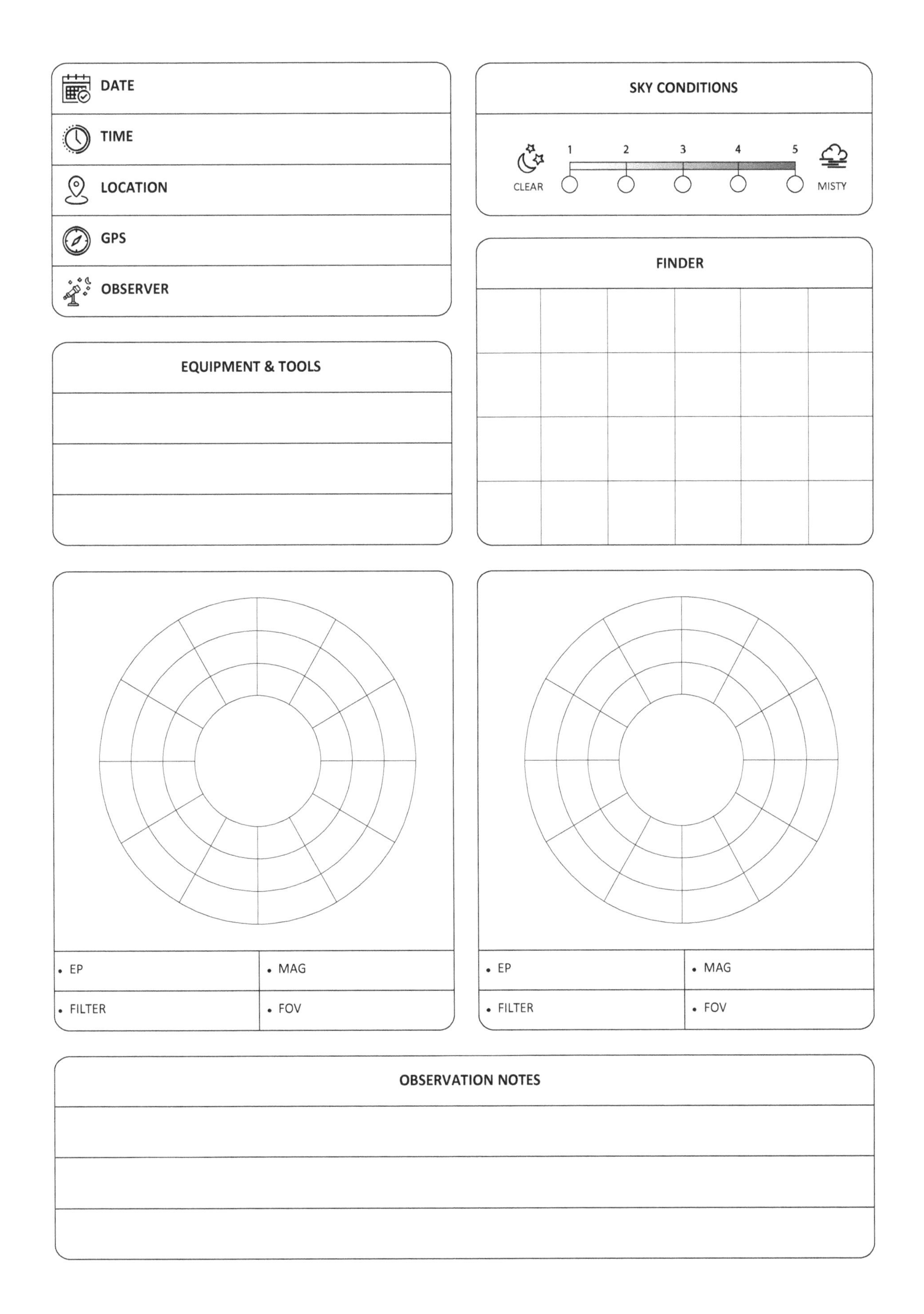
DATE
TIME
LOCATION
GPS
OBSERVER
SKY CONDITIONS
1
2
3
4
5
CLEAR
MISTY
FINDER
EQUIPMENT & TOOLS
EP
MAG
FILTER
FOV
EP
MAG
FILTER
FOV
OBSERVATION NOTES

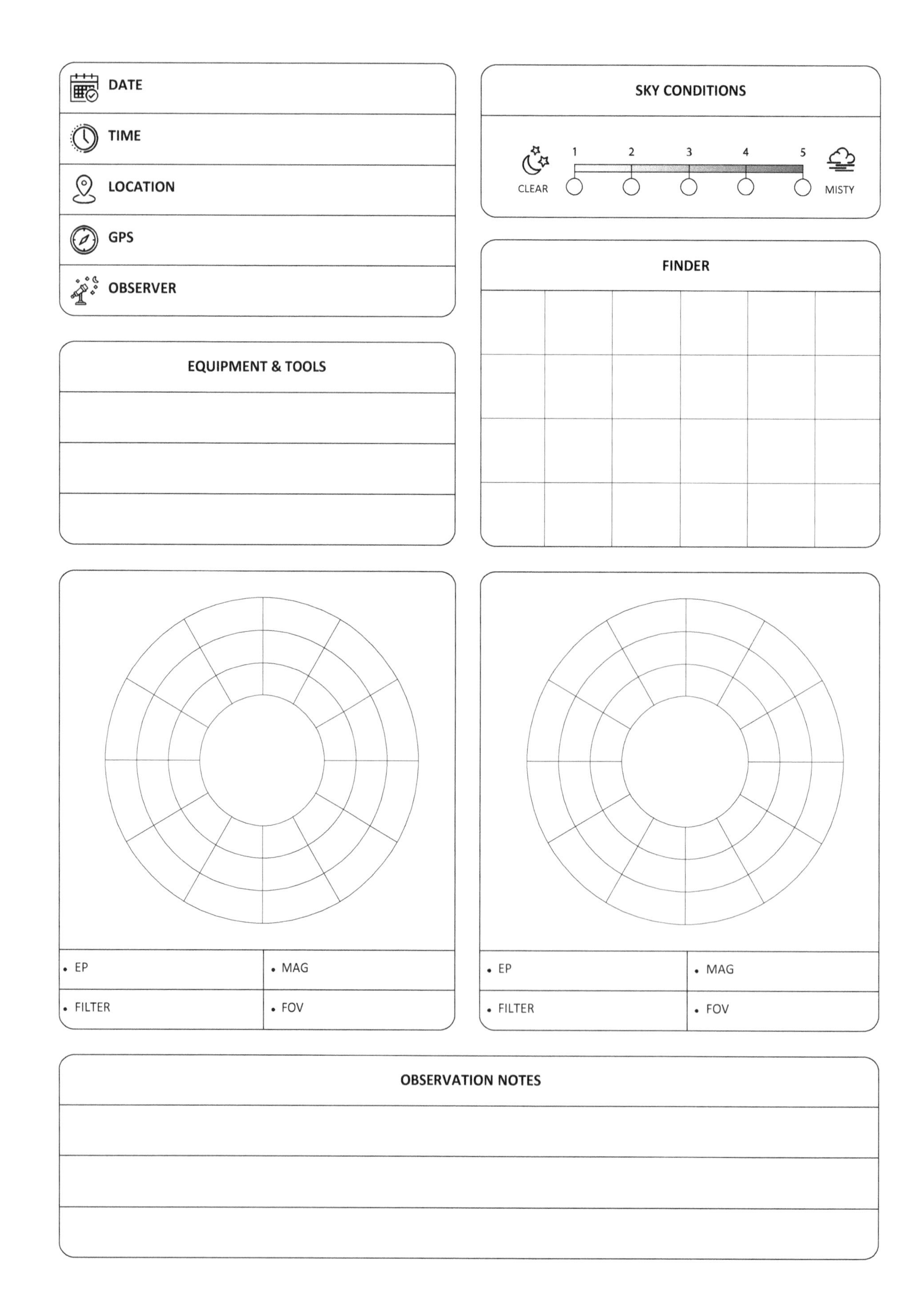

DATE
TIME
LOCATION
GPS
OBSERVER
SKY CONDITIONS
1
2
3
4
5
CLEAR
MISTY
FINDER
EQUIPMENT & TOOLS
EP
MAG
FILTER
FOV
EP
MAG
FILTER
FOV
OBSERVATION NOTES

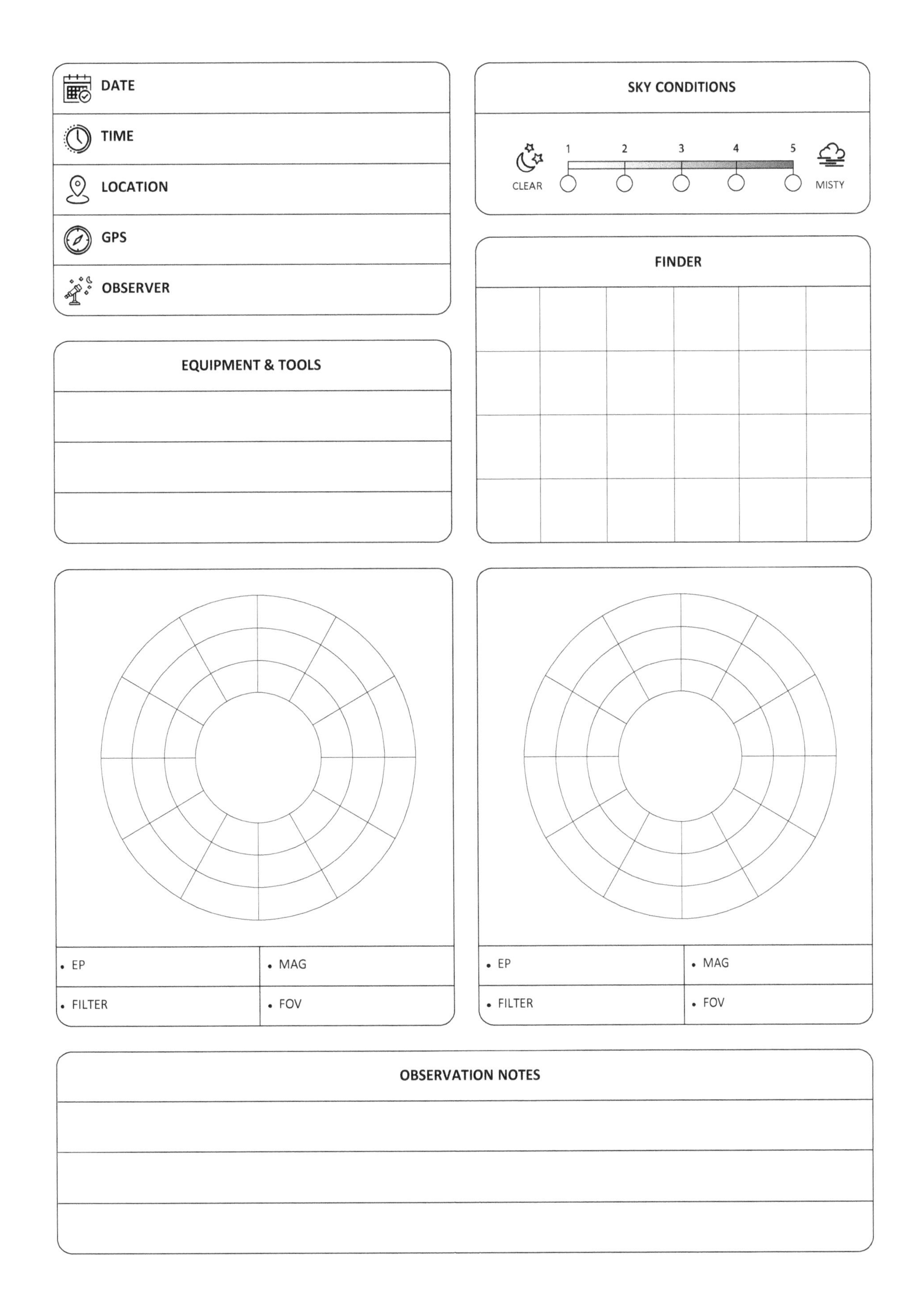
DATE
TIME
LOCATION
GPS
OBSERVER
SKY CONDITIONS
1
2
3
4
5
CLEAR
MISTY
FINDER
EQUIPMENT & TOOLS
EP
MAG
FILTER
FOV
EP
MAG
FILTER
FOV
OBSERVATION NOTES

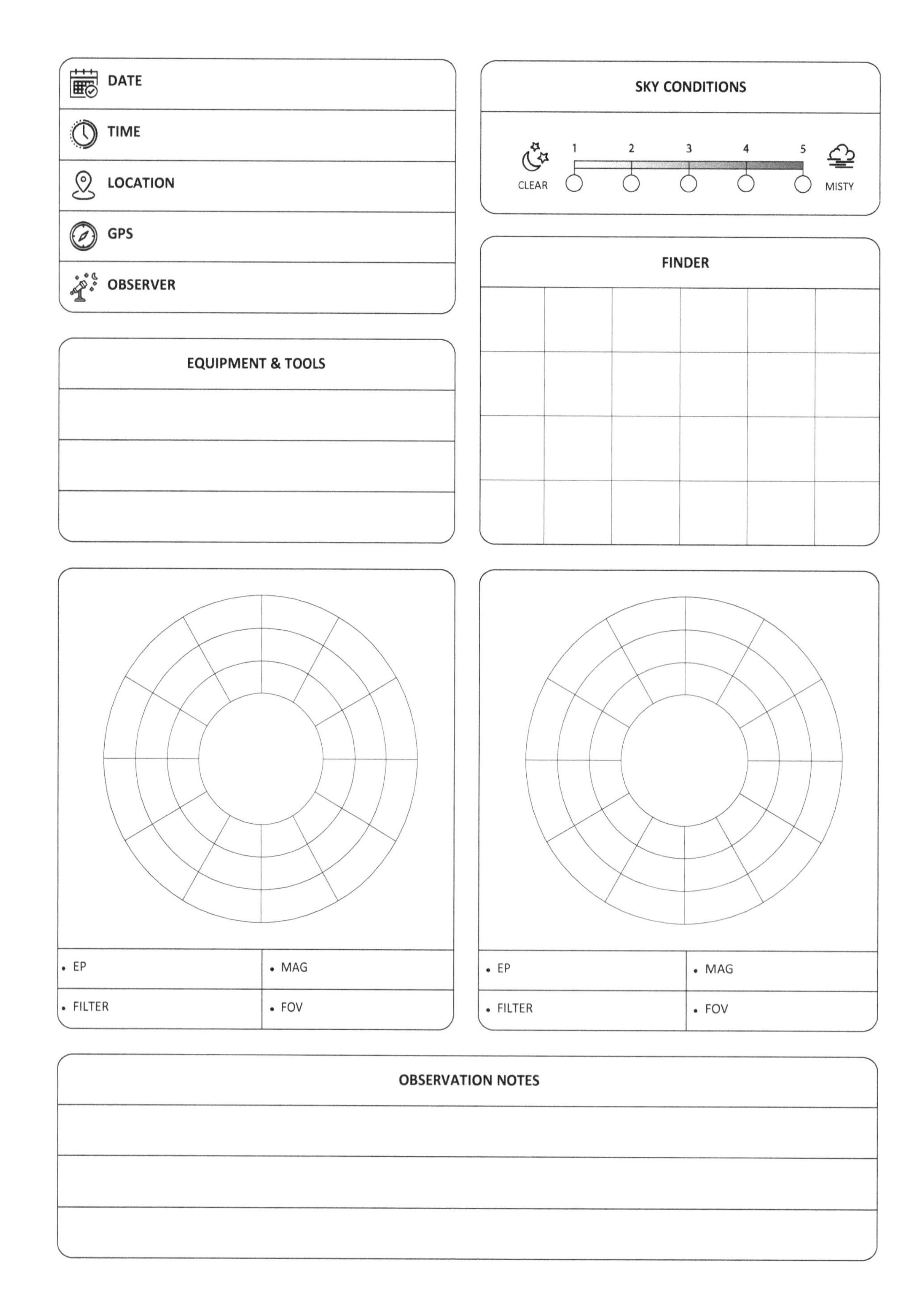
DATE
TIME
LOCATION
GPS
OBSERVER
SKY CONDITIONS
1
2
3
4
5
CLEAR
MISTY
EQUIPMENT & TOOLS
FINDER
EP
MAG
FILTER
FOV
EP
MAG
FILTER
FOV
OBSERVATION NOTES

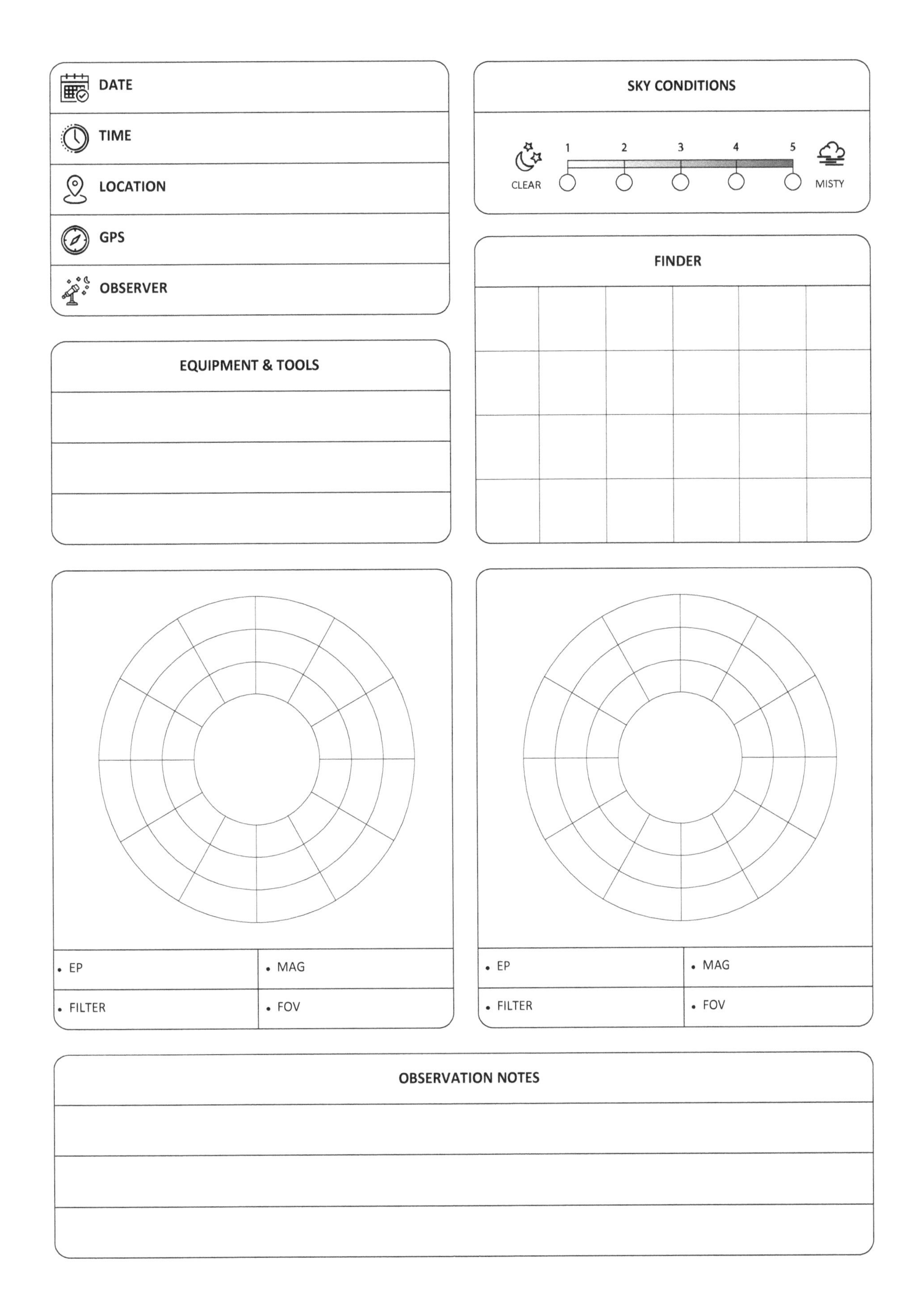
DATE
TIME
LOCATION
GPS
OBSERVER
SKY CONDITIONS
1
2
3
4
5
CLEAR
MISTY
FINDER
EQUIPMENT & TOOLS
• EP
• MAG
• FILTER
• FOV
• EP
• MAG
• FILTER
• FOV
OBSERVATION NOTES

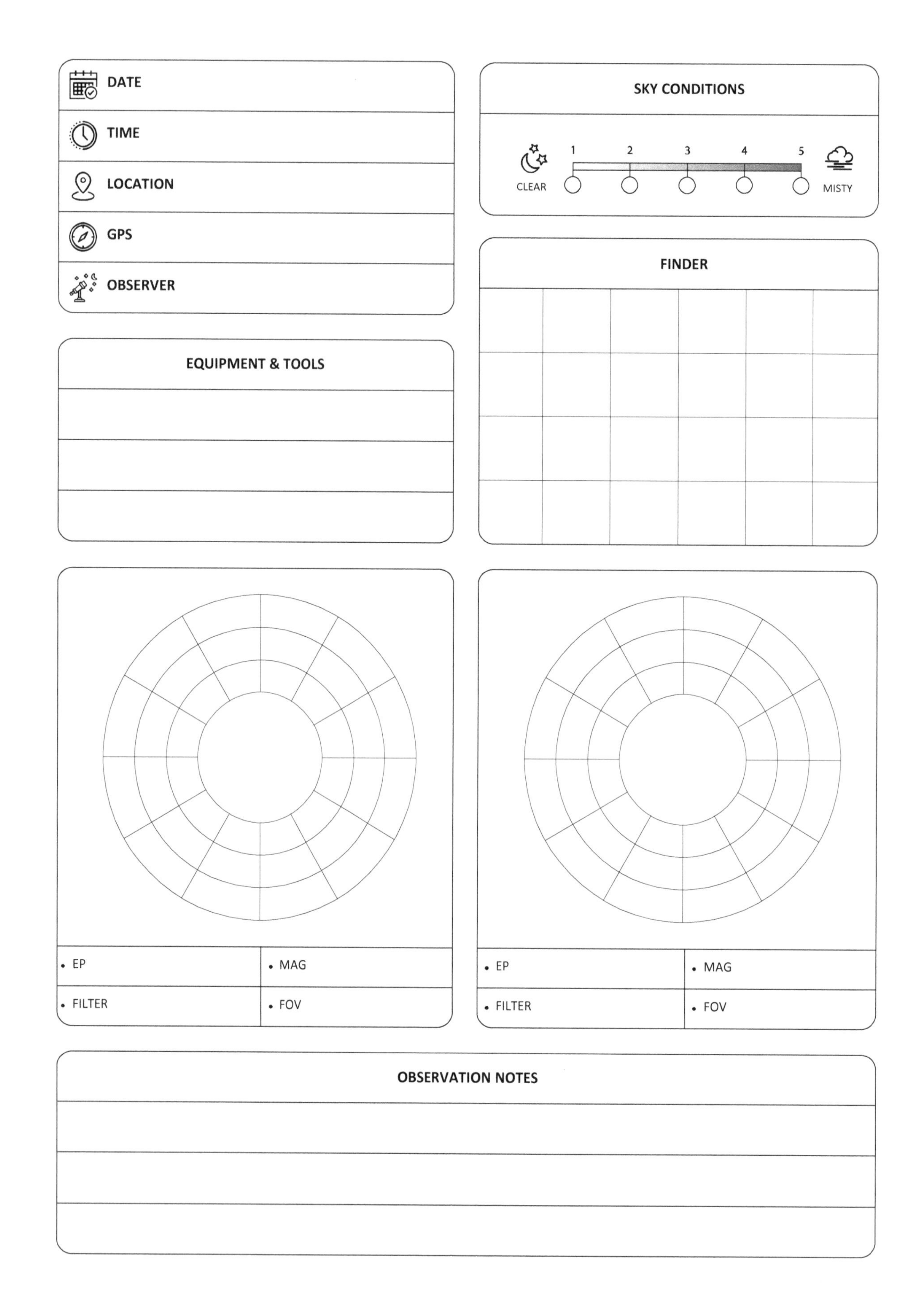

DATE

TIME

LOCATION

GPS

OBSERVER

SKY CONDITIONS

CLEAR 1 2 3 4 5 MISTY

FINDER

EQUIPMENT & TOOLS

- EP
- MAG
- FILTER
- FOV

- EP
- MAG
- FILTER
- FOV

OBSERVATION NOTES

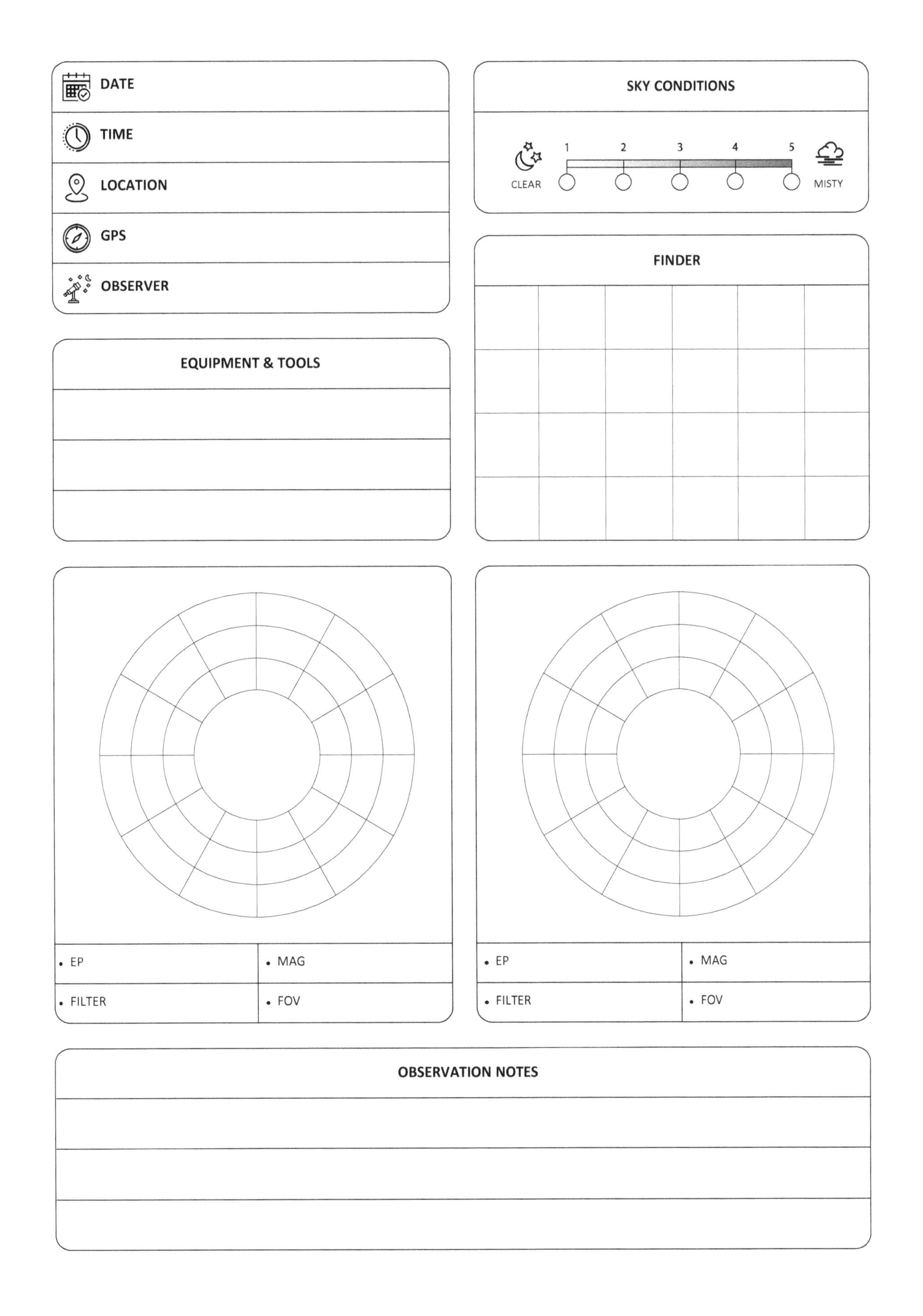
DATE
TIME
LOCATION
GPS
OBSERVER
SKY CONDITIONS
1
2
3
4
5
CLEAR
MISTY
EQUIPMENT & TOOLS
FINDER
EP
MAG
FILTER
FOV
EP
MAG
FILTER
FOV
OBSERVATION NOTES

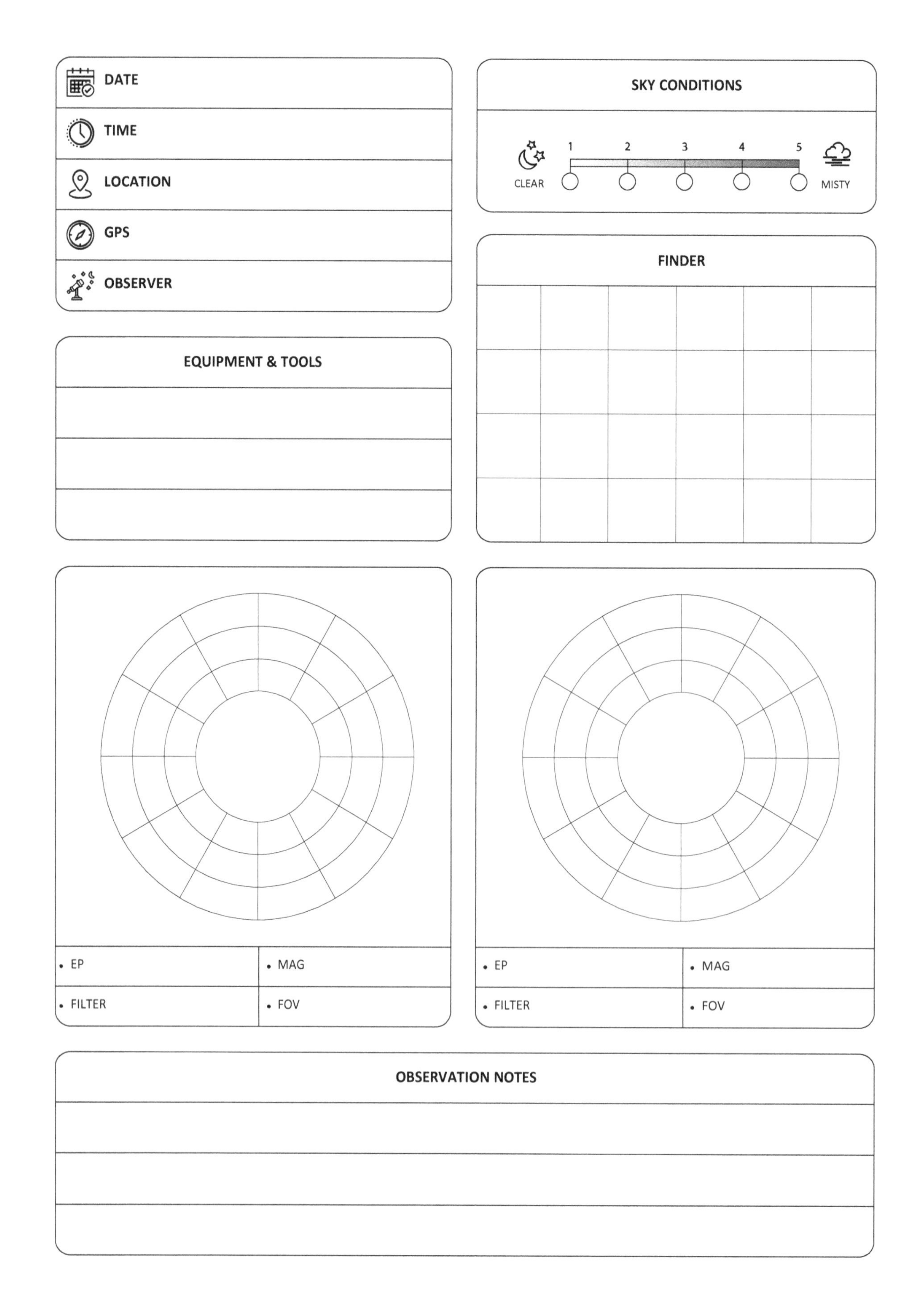
DATE
TIME
LOCATION
GPS
OBSERVER
SKY CONDITIONS
1
2
3
4
5
CLEAR
MISTY
FINDER
EQUIPMENT & TOOLS
EP
MAG
FILTER
FOV
EP
MAG
FILTER
FOV
OBSERVATION NOTES

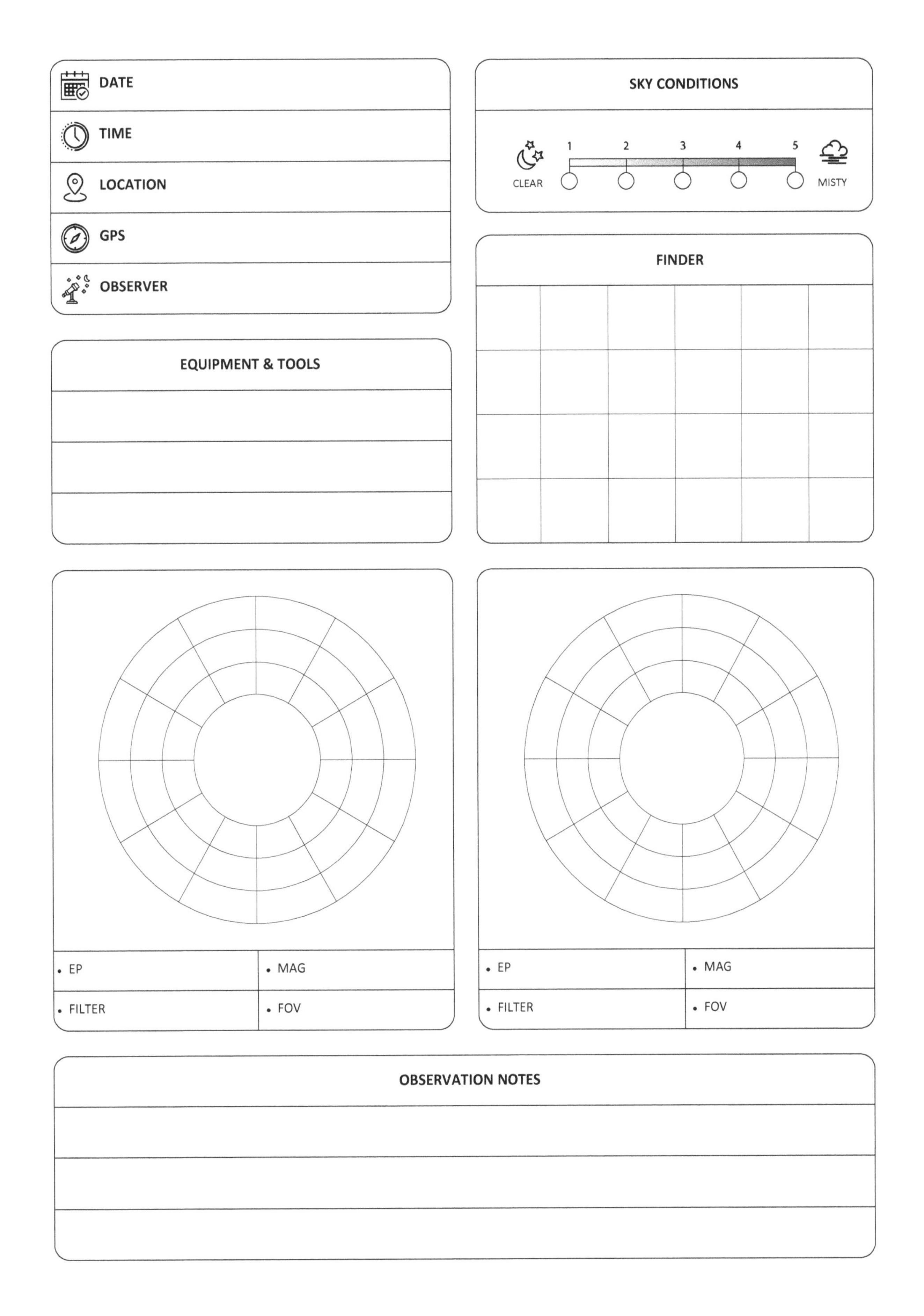
DATE
TIME
LOCATION
GPS
OBSERVER
SKY CONDITIONS
1
2
3
4
5
CLEAR
MISTY
FINDER
EQUIPMENT & TOOLS
EP
MAG
FILTER
FOV
EP
MAG
FILTER
FOV
OBSERVATION NOTES

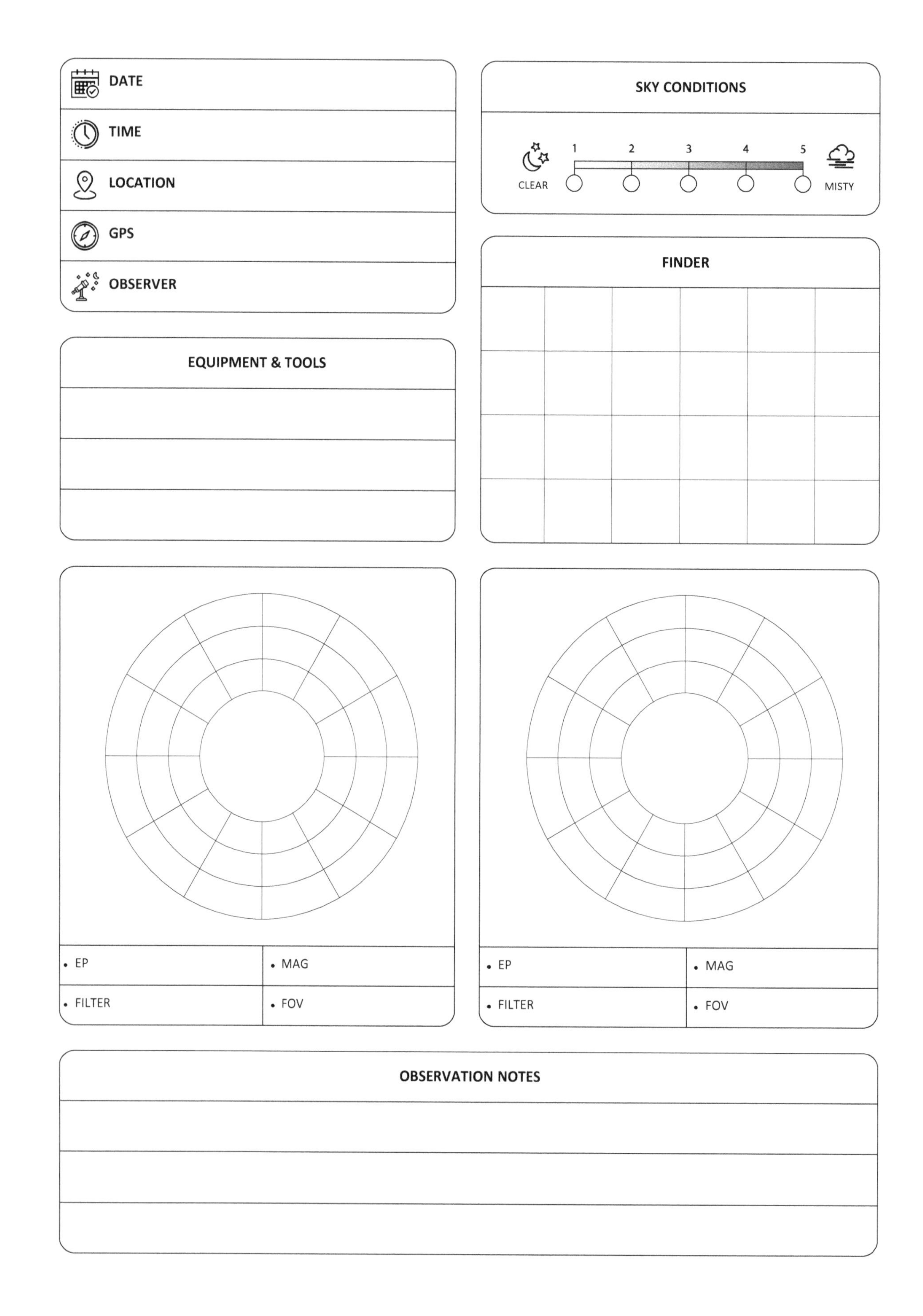
DATE
TIME
LOCATION
GPS
OBSERVER
SKY CONDITIONS
1
2
3
4
5
CLEAR
MISTY
FINDER
EQUIPMENT & TOOLS
EP
MAG
FILTER
FOV
EP
MAG
FILTER
FOV
OBSERVATION NOTES

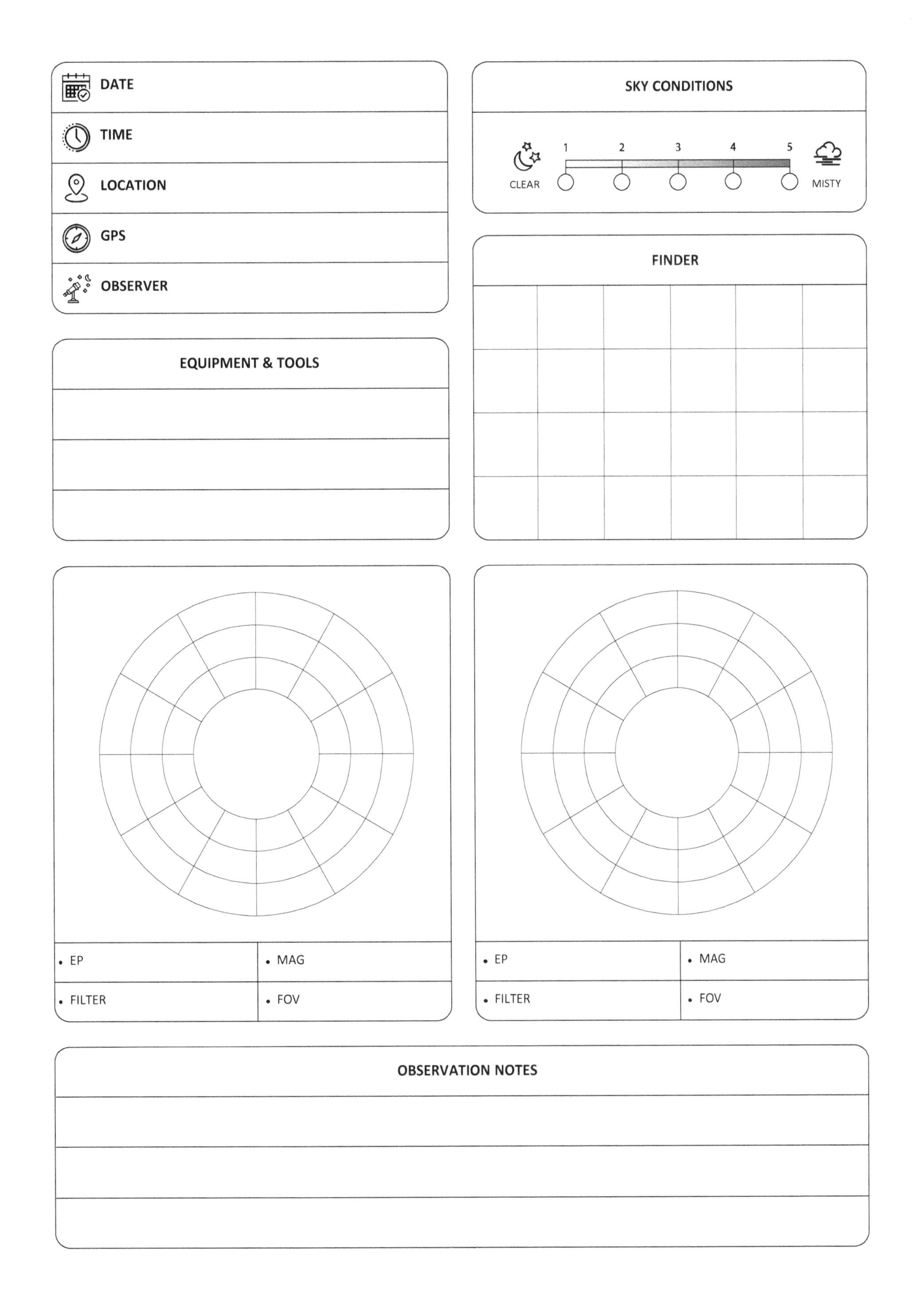
DATE
TIME
LOCATION
GPS
OBSERVER
SKY CONDITIONS
1
2
3
4
5
CLEAR
MISTY
FINDER
EQUIPMENT & TOOLS
EP
MAG
FILTER
FOV
EP
MAG
FILTER
FOV
OBSERVATION NOTES

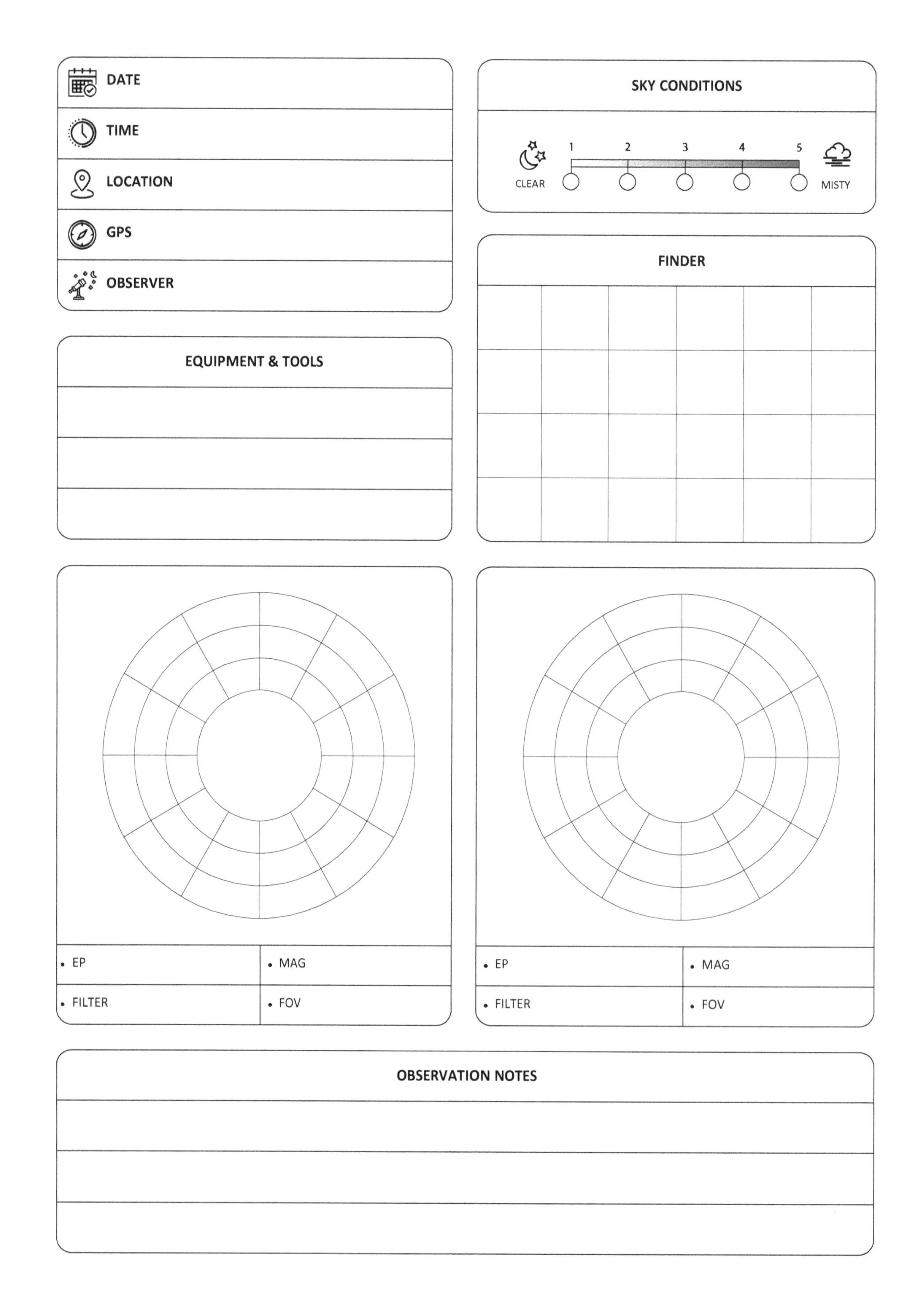
DATE
TIME
LOCATION
GPS
OBSERVER
SKY CONDITIONS
CLEAR
1
2
3
4
5
MISTY
FINDER
EQUIPMENT & TOOLS
EP
MAG
FILTER
FOV
EP
MAG
FILTER
FOV
OBSERVATION NOTES

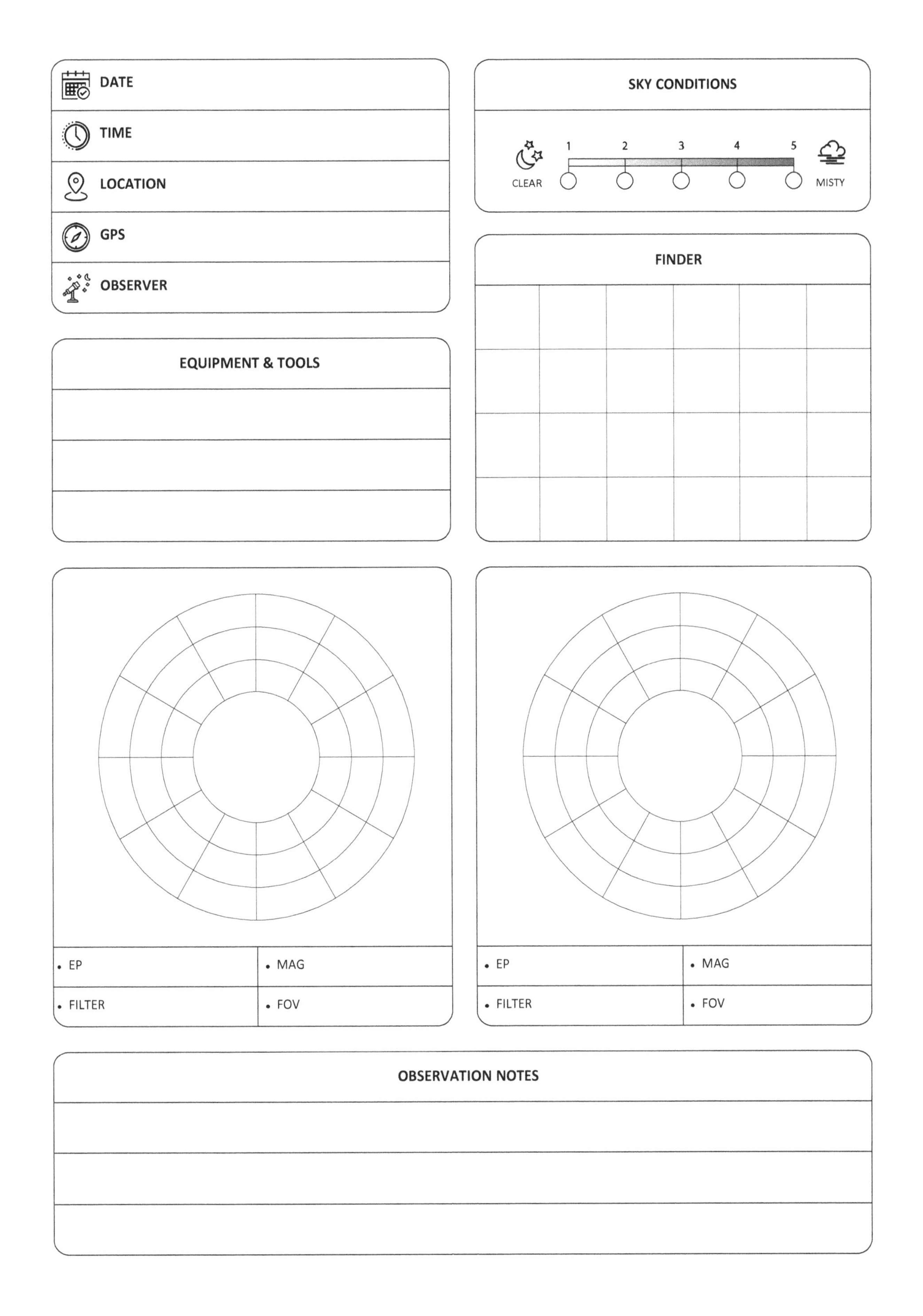
DATE
TIME
LOCATION
GPS
OBSERVER
SKY CONDITIONS
1
2
3
4
5
CLEAR
MISTY
FINDER
EQUIPMENT & TOOLS
EP
MAG
FILTER
FOV
EP
MAG
FILTER
FOV
OBSERVATION NOTES

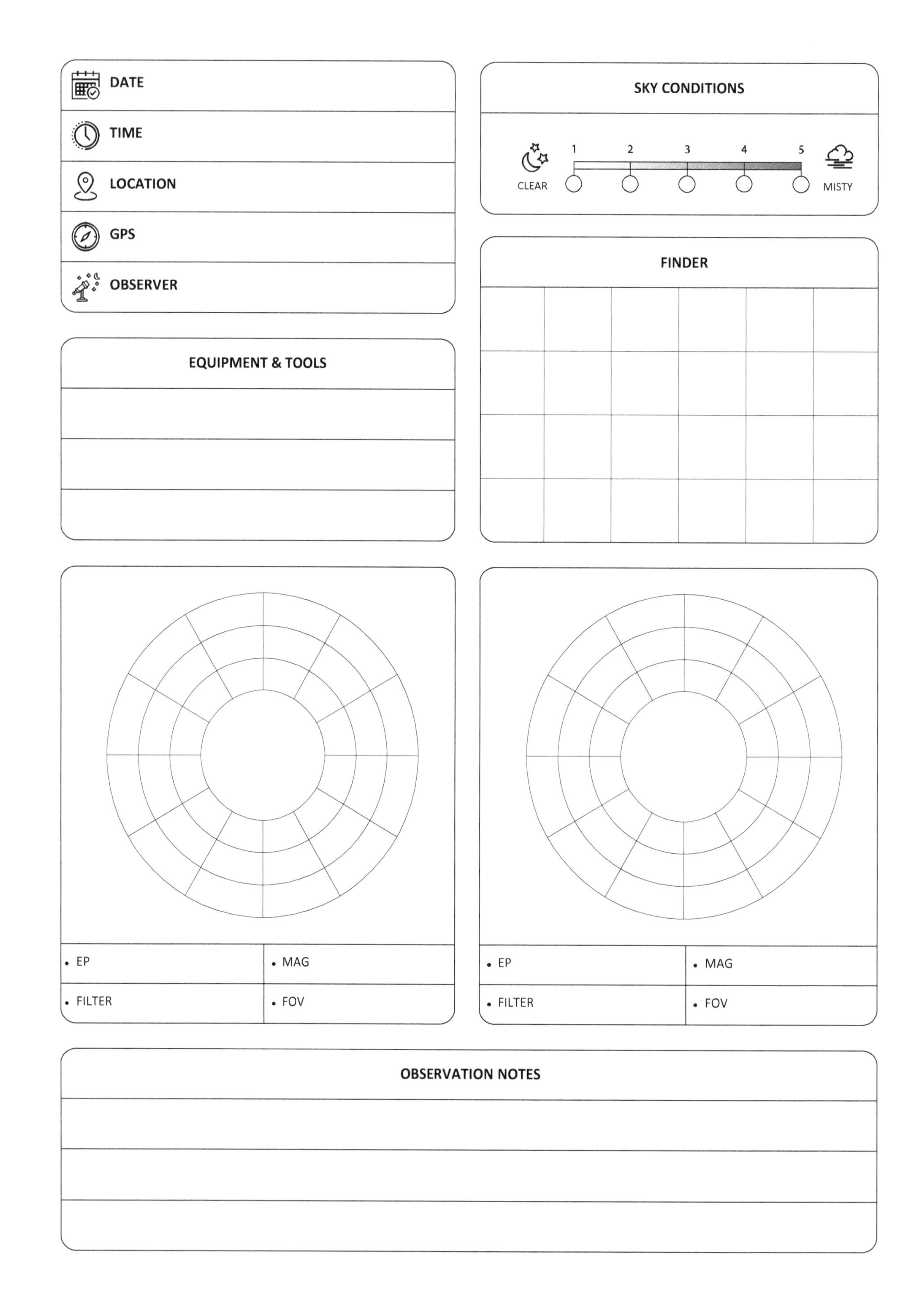
DATE
TIME
LOCATION
GPS
OBSERVER
SKY CONDITIONS
1
2
3
4
5
CLEAR
MISTY
FINDER
EQUIPMENT & TOOLS
EP
MAG
FILTER
FOV
EP
MAG
FILTER
FOV
OBSERVATION NOTES

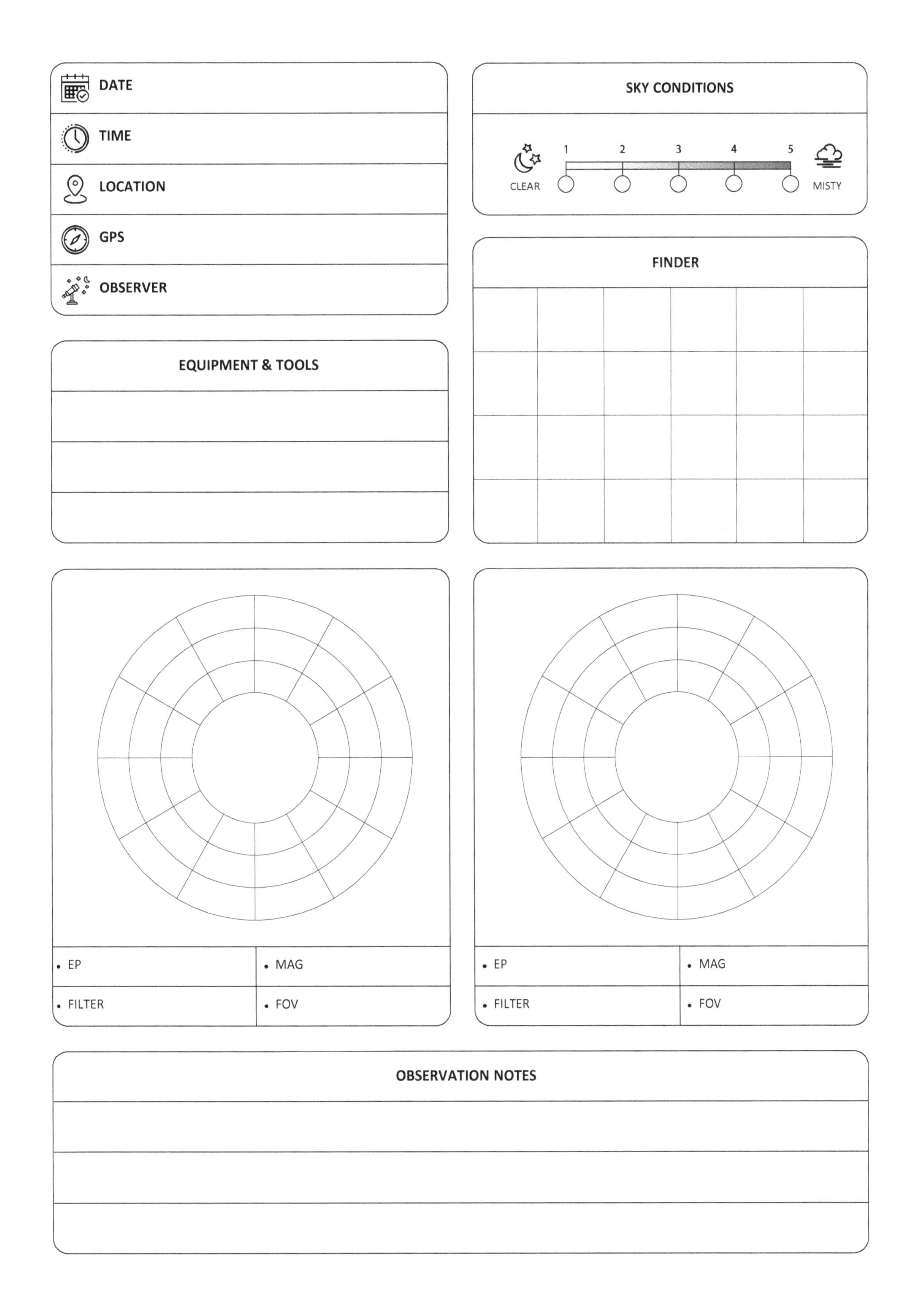

DATE
TIME
LOCATION
GPS
OBSERVER
SKY CONDITIONS
1
2
3
4
5
CLEAR
MISTY
EQUIPMENT & TOOLS
FINDER
EP
MAG
FILTER
FOV
EP
MAG
FILTER
FOV
OBSERVATION NOTES

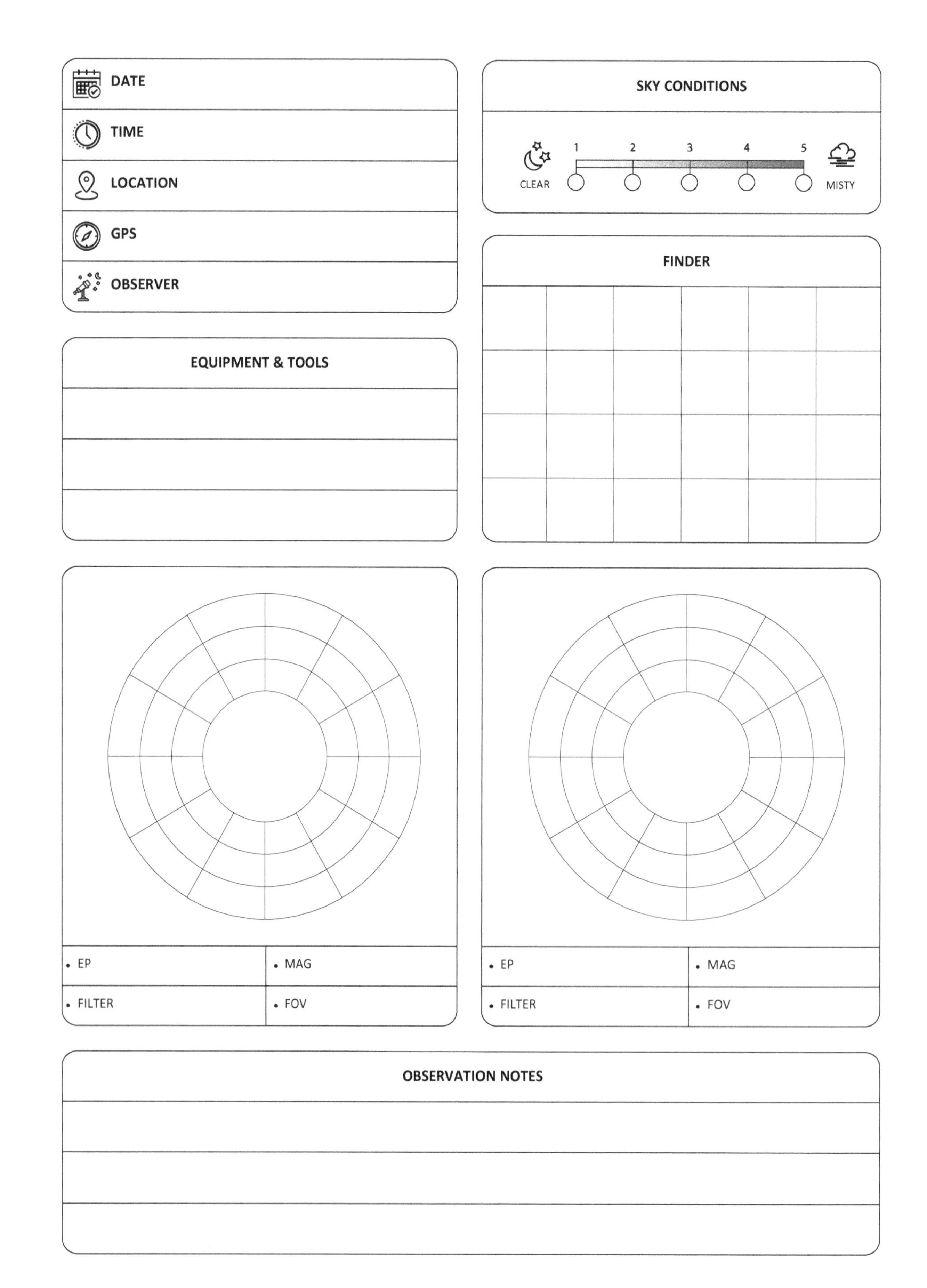
DATE
TIME
LOCATION
GPS
OBSERVER
SKY CONDITIONS
1
2
3
4
5
CLEAR
MISTY
FINDER
EQUIPMENT & TOOLS
EP
MAG
FILTER
FOV
EP
MAG
FILTER
FOV
OBSERVATION NOTES

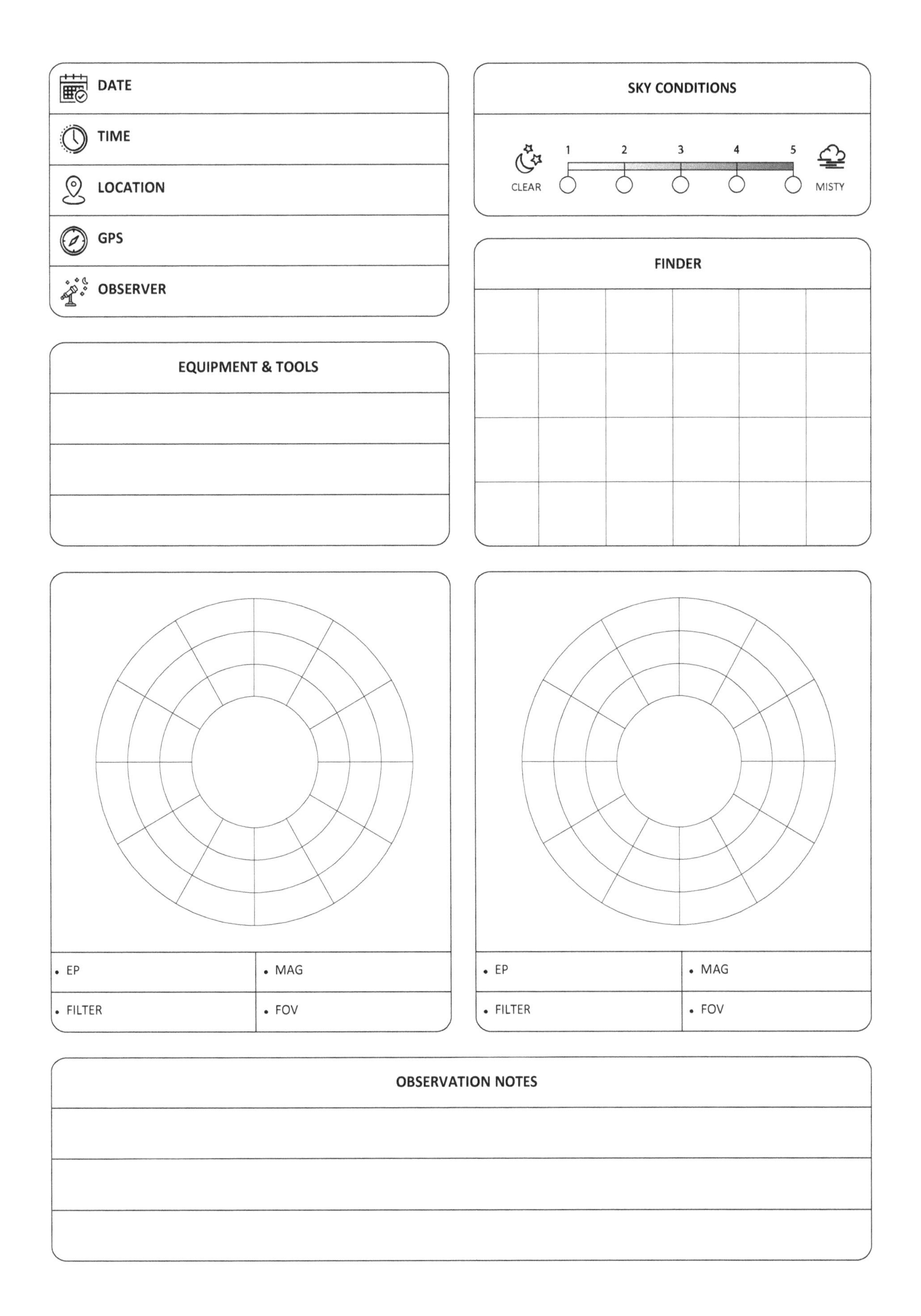
DATE
TIME
LOCATION
GPS
OBSERVER
SKY CONDITIONS
1
2
3
4
5
CLEAR
MISTY
FINDER
EQUIPMENT & TOOLS
EP
MAG
FILTER
FOV
EP
MAG
FILTER
FOV
OBSERVATION NOTES

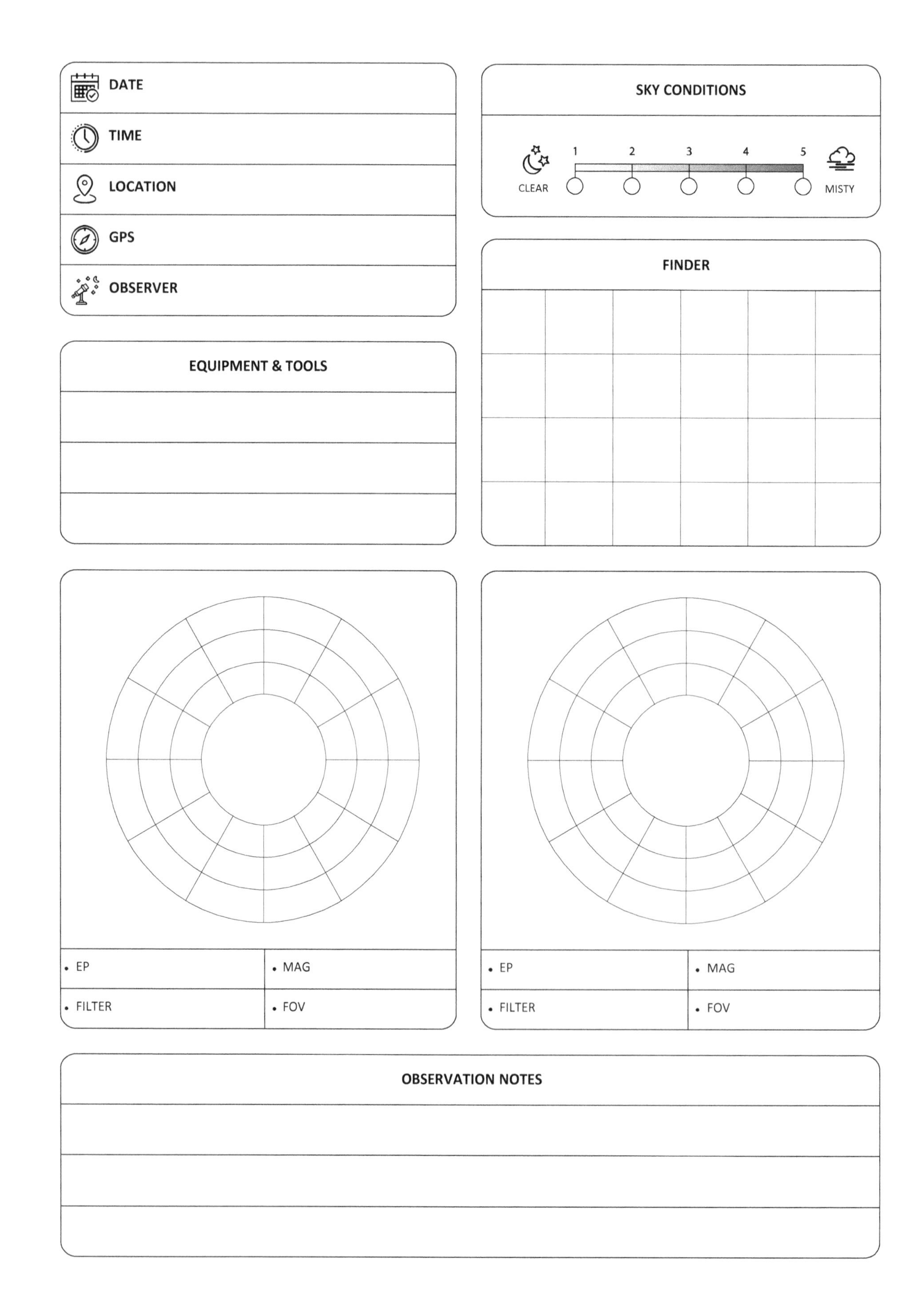
DATE
TIME
LOCATION
GPS
OBSERVER
SKY CONDITIONS
1
2
3
4
5
CLEAR
MISTY
FINDER
EQUIPMENT & TOOLS
EP
MAG
FILTER
FOV
EP
MAG
FILTER
FOV
OBSERVATION NOTES

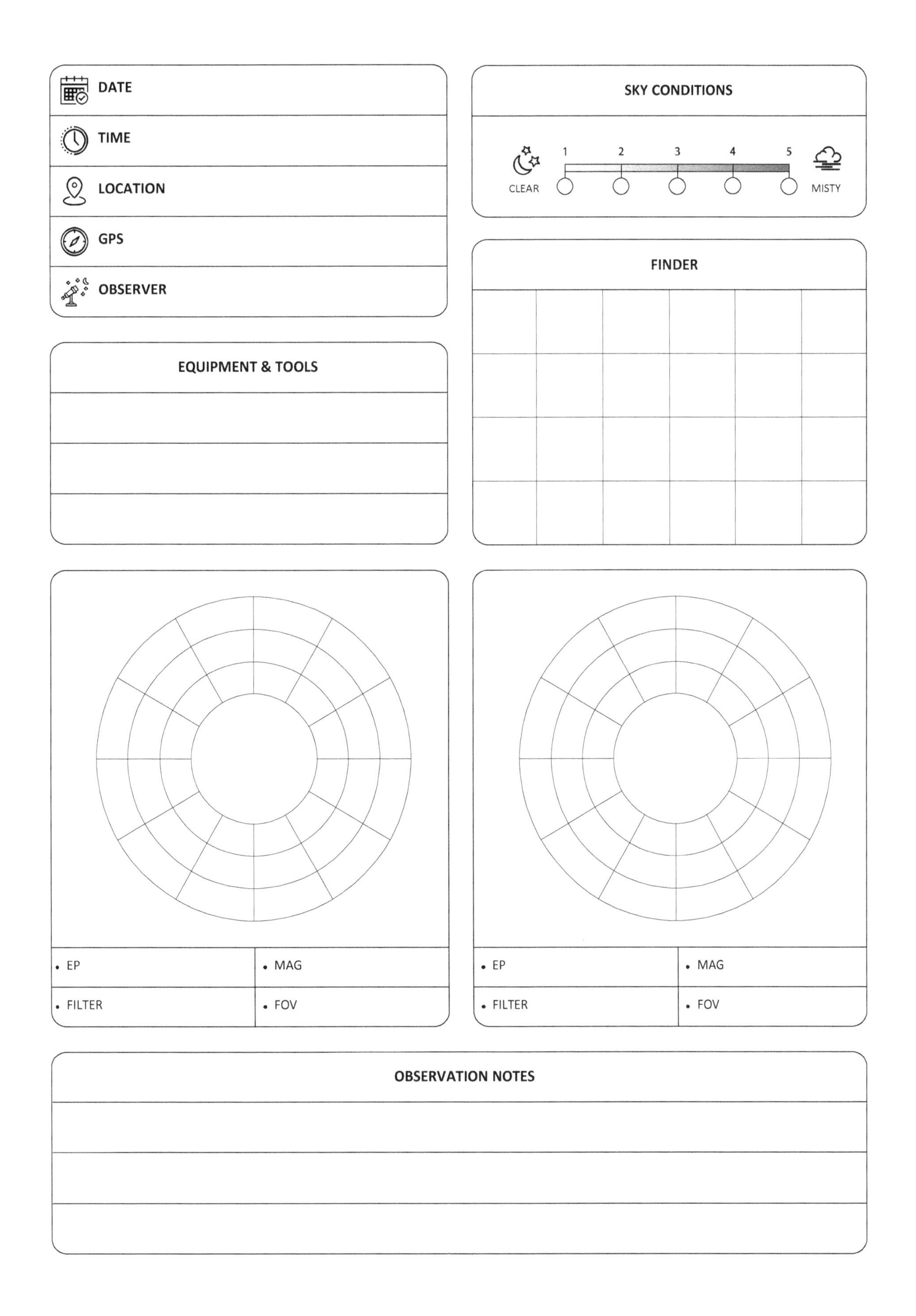

DATE

TIME

LOCATION

GPS

OBSERVER

SKY CONDITIONS

CLEAR 1 2 3 4 5 MISTY

FINDER

EQUIPMENT & TOOLS

• EP • MAG
• FILTER • FOV

• EP • MAG
• FILTER • FOV

OBSERVATION NOTES

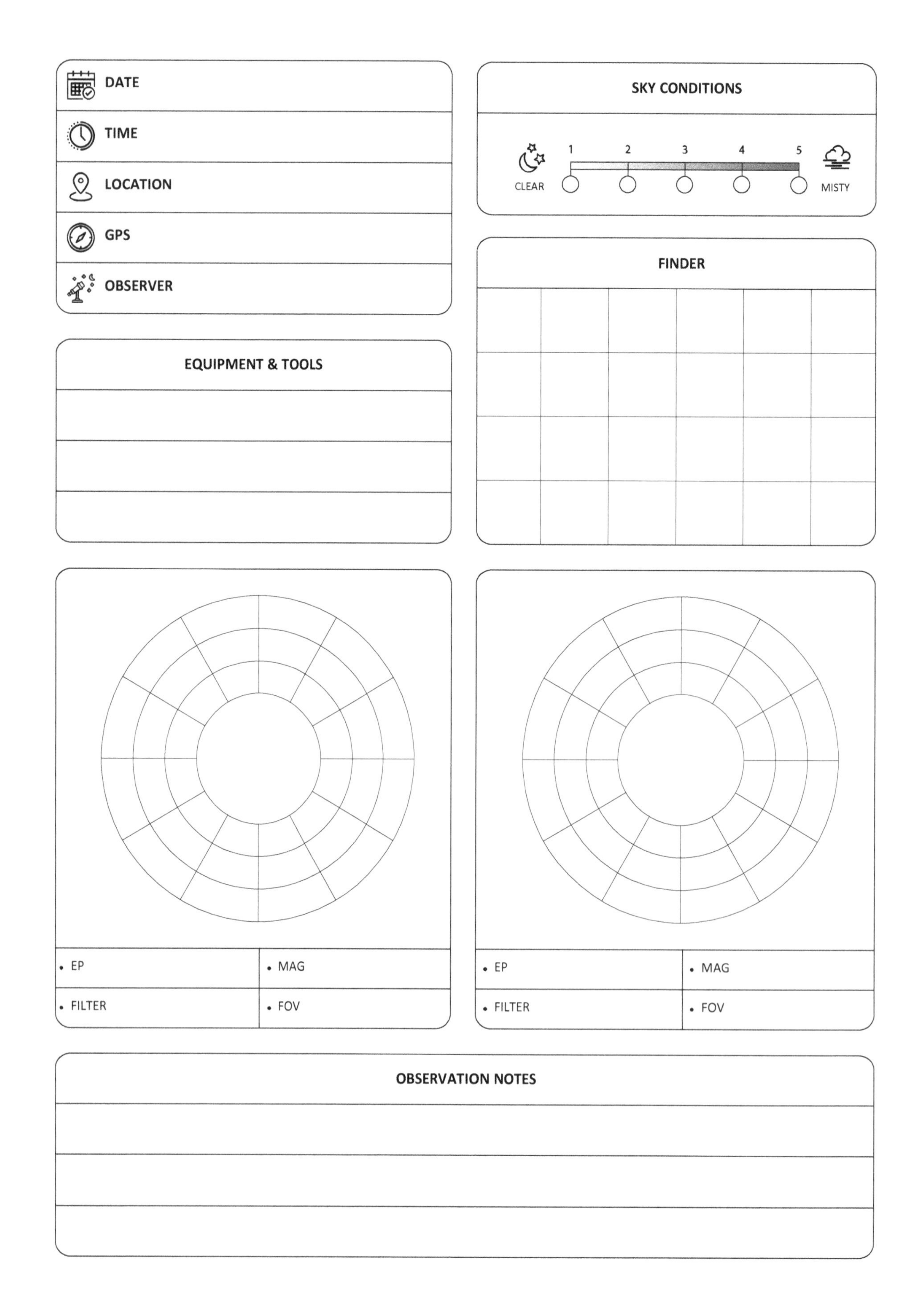
DATE
TIME
LOCATION
GPS
OBSERVER
SKY CONDITIONS
1
2
3
4
5
CLEAR
MISTY
FINDER
EQUIPMENT & TOOLS
EP
MAG
FILTER
FOV
EP
MAG
FILTER
FOV
OBSERVATION NOTES

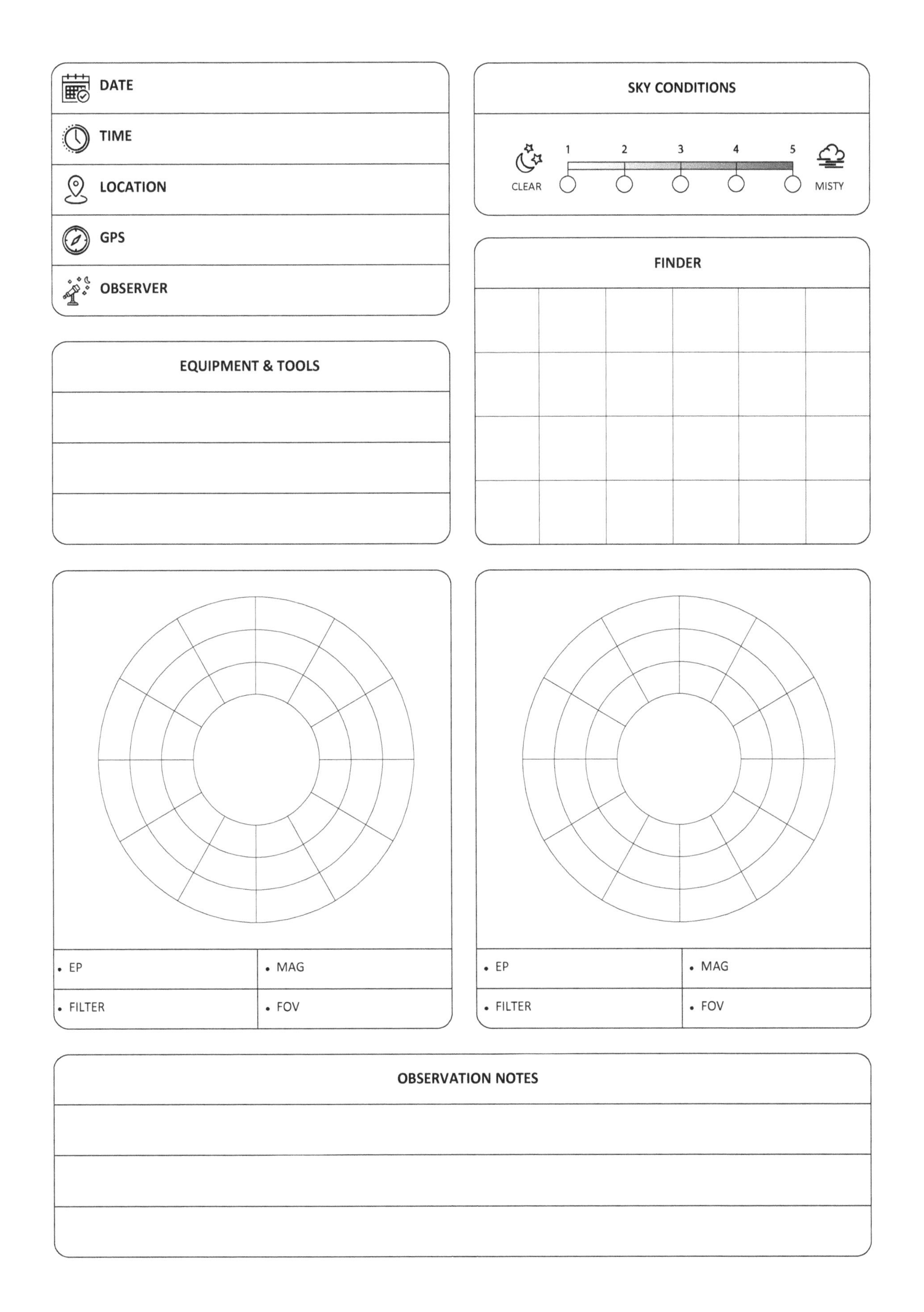
DATE
TIME
LOCATION
GPS
OBSERVER
SKY CONDITIONS
1
2
3
4
5
CLEAR
MISTY
EQUIPMENT & TOOLS
FINDER
• EP
• MAG
• FILTER
• FOV
• EP
• MAG
• FILTER
• FOV
OBSERVATION NOTES

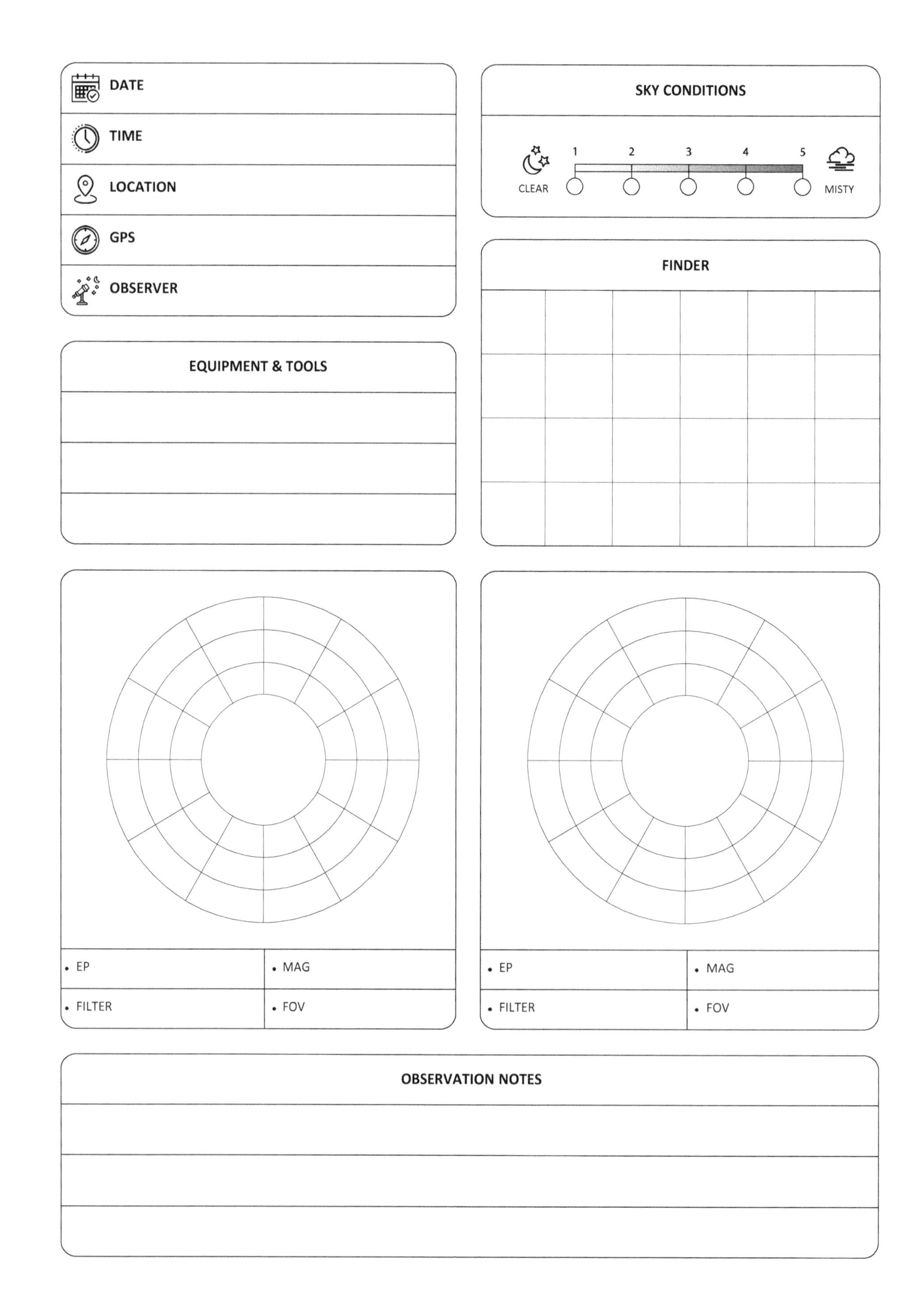
DATE
TIME
LOCATION
GPS
OBSERVER
SKY CONDITIONS
1
2
3
4
5
CLEAR
MISTY
EQUIPMENT & TOOLS
FINDER
EP
MAG
FILTER
FOV
EP
MAG
FILTER
FOV
OBSERVATION NOTES

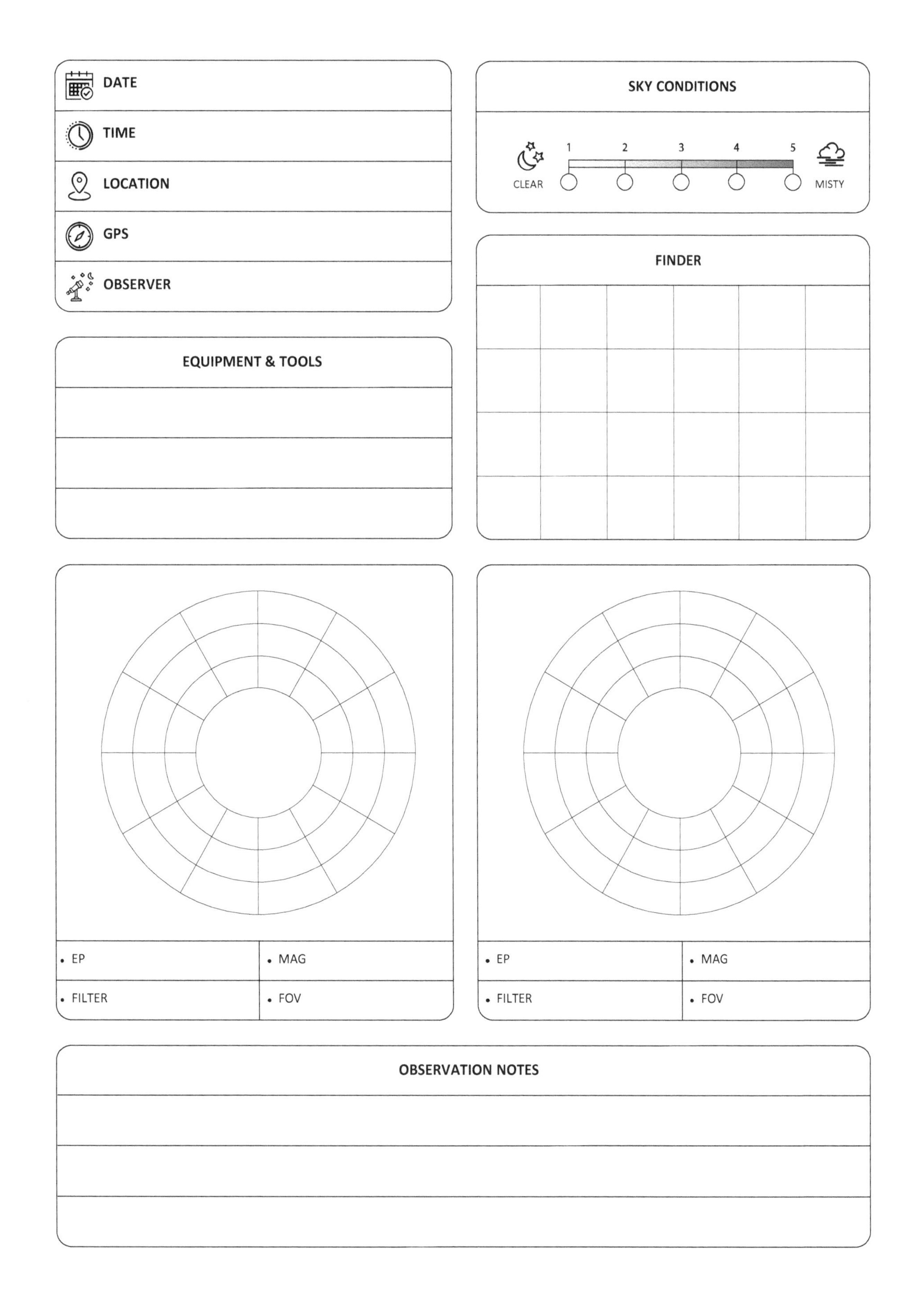
DATE
TIME
LOCATION
GPS
OBSERVER
SKY CONDITIONS
CLEAR
1
2
3
4
5
MISTY
FINDER
EQUIPMENT & TOOLS
EP
MAG
FILTER
FOV
EP
MAG
FILTER
FOV
OBSERVATION NOTES

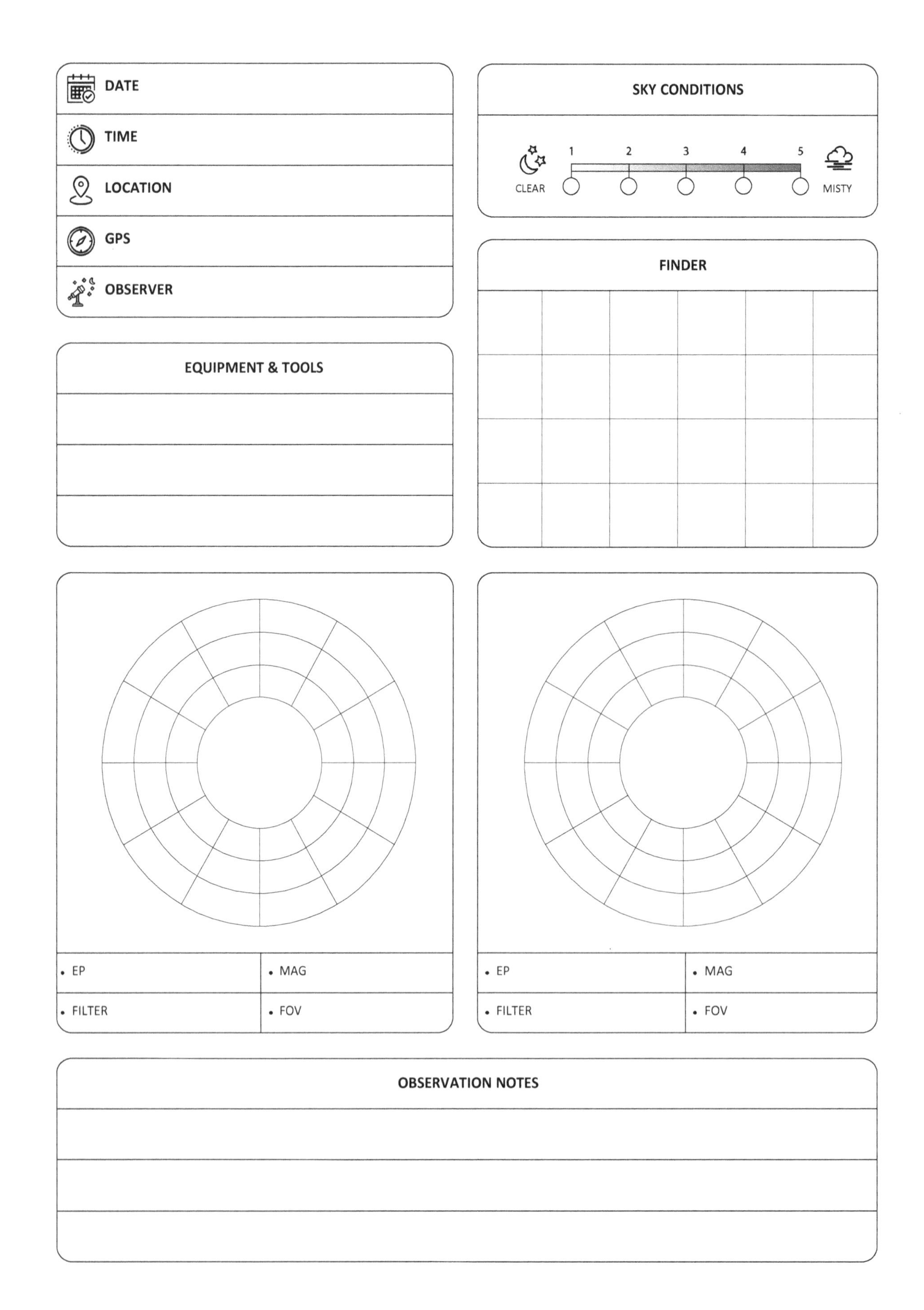
DATE
TIME
LOCATION
GPS
OBSERVER
SKY CONDITIONS
1
2
3
4
5
CLEAR
MISTY
FINDER
EQUIPMENT & TOOLS
• EP
• MAG
• FILTER
• FOV
• EP
• MAG
• FILTER
• FOV
OBSERVATION NOTES

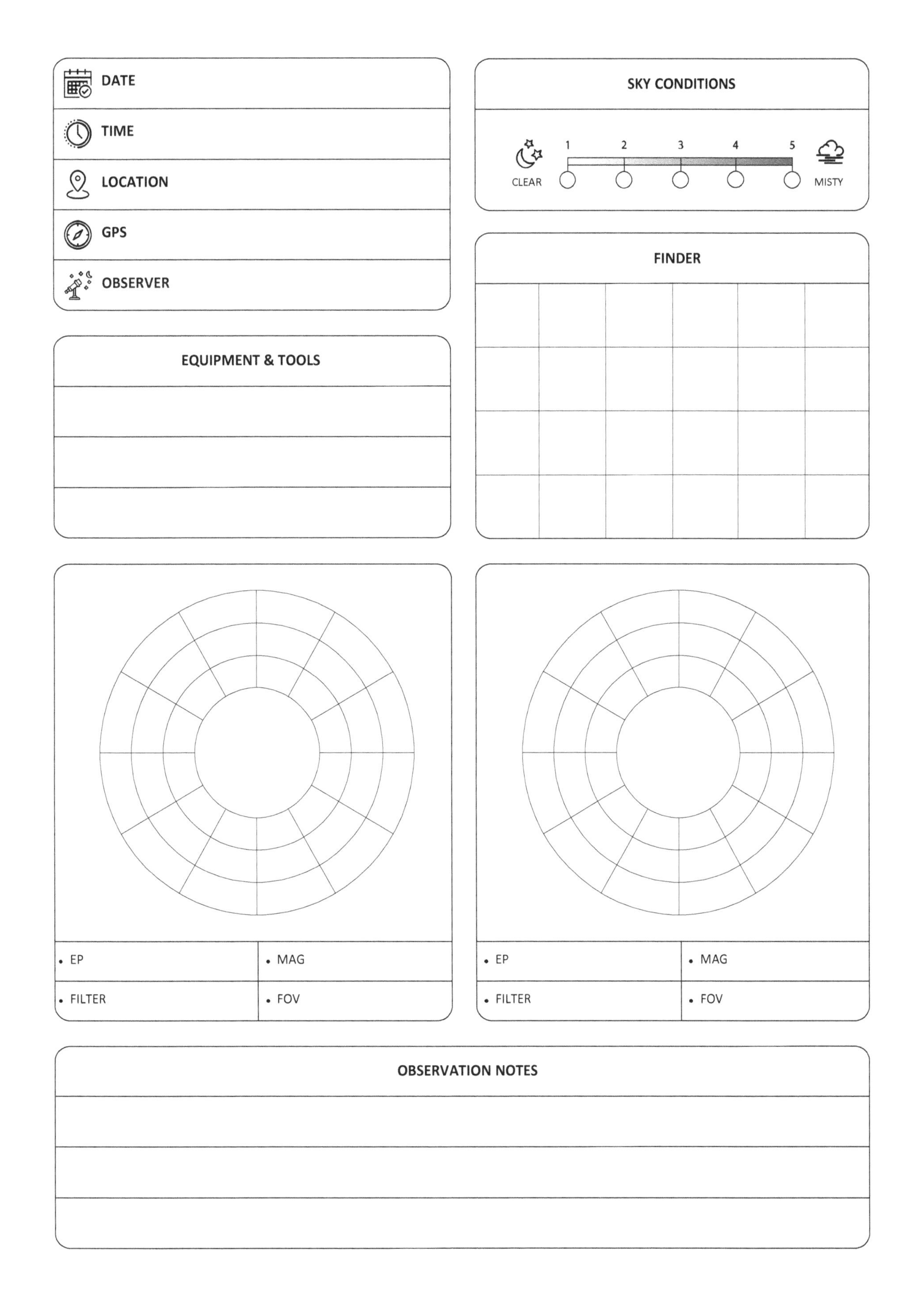

DATE
TIME
LOCATION
GPS
OBSERVER
SKY CONDITIONS
1
2
3
4
5
CLEAR
MISTY
EQUIPMENT & TOOLS
FINDER
EP
MAG
FILTER
FOV
EP
MAG
FILTER
FOV
OBSERVATION NOTES

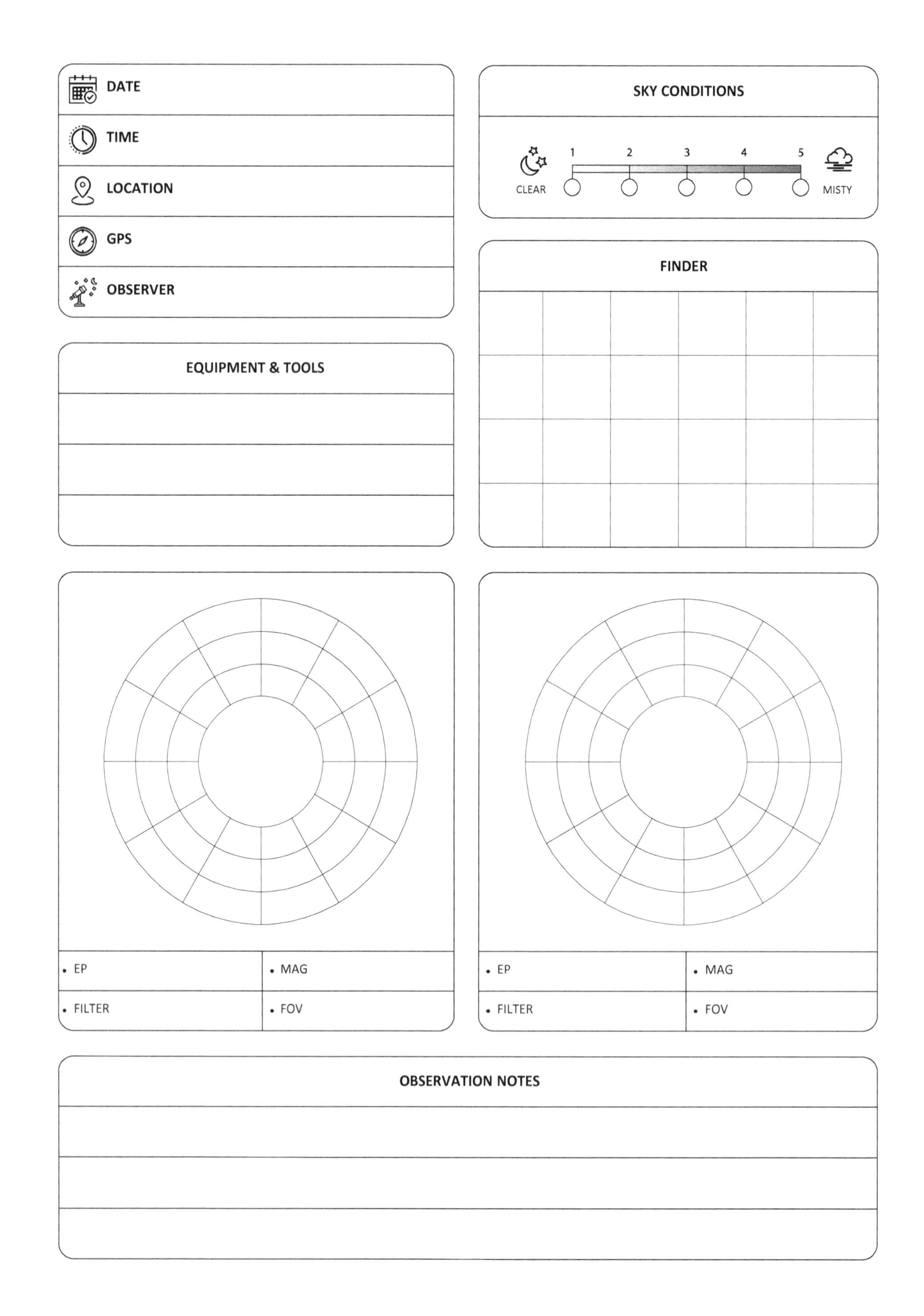
DATE
TIME
LOCATION
GPS
OBSERVER
SKY CONDITIONS
1
2
3
4
5
CLEAR
MISTY
FINDER
EQUIPMENT & TOOLS
EP
MAG
FILTER
FOV
EP
MAG
FILTER
FOV
OBSERVATION NOTES

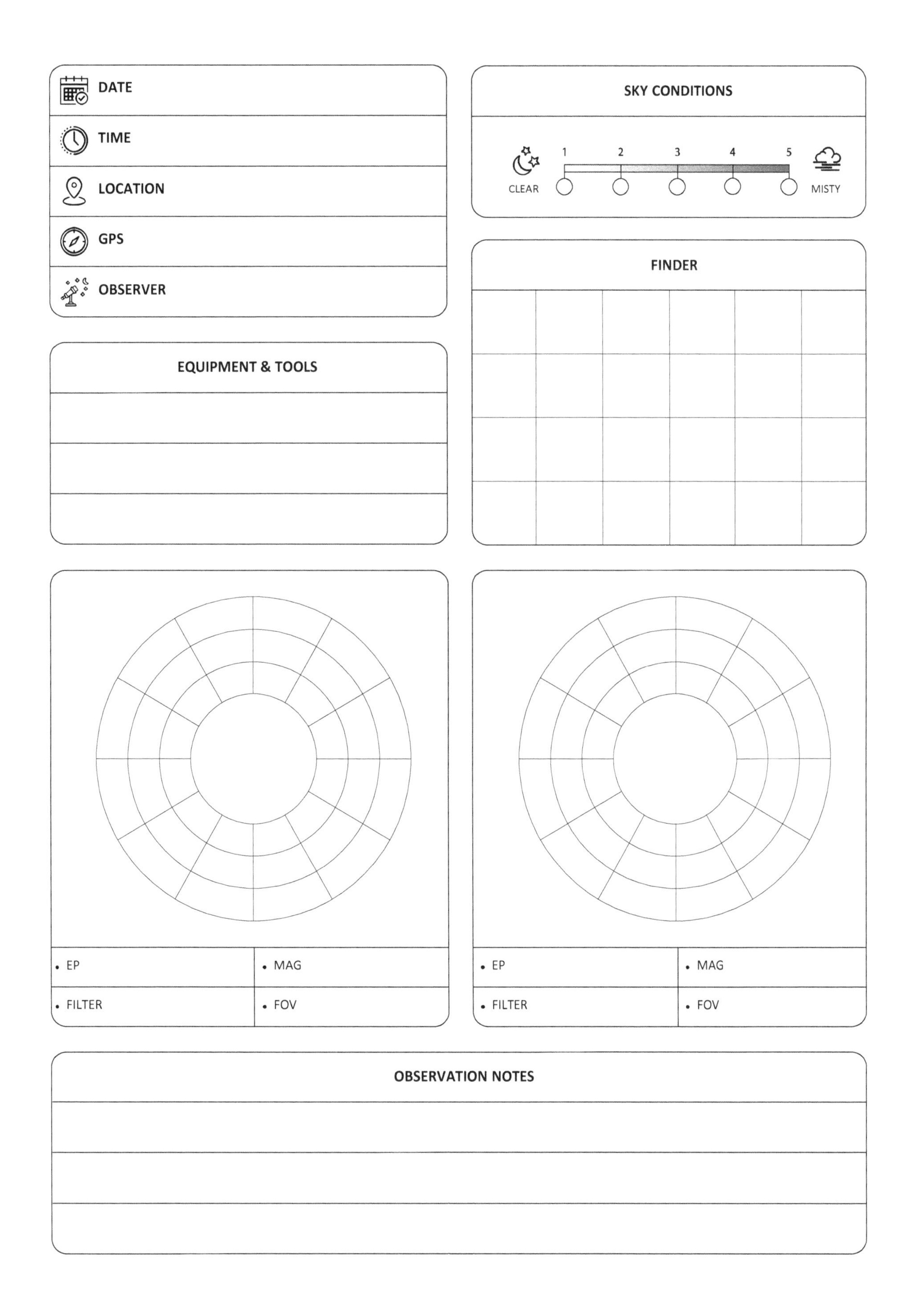
DATE
TIME
LOCATION
GPS
OBSERVER
SKY CONDITIONS
CLEAR
1
2
3
4
5
MISTY
EQUIPMENT & TOOLS
FINDER
EP
MAG
FILTER
FOV
EP
MAG
FILTER
FOV
OBSERVATION NOTES

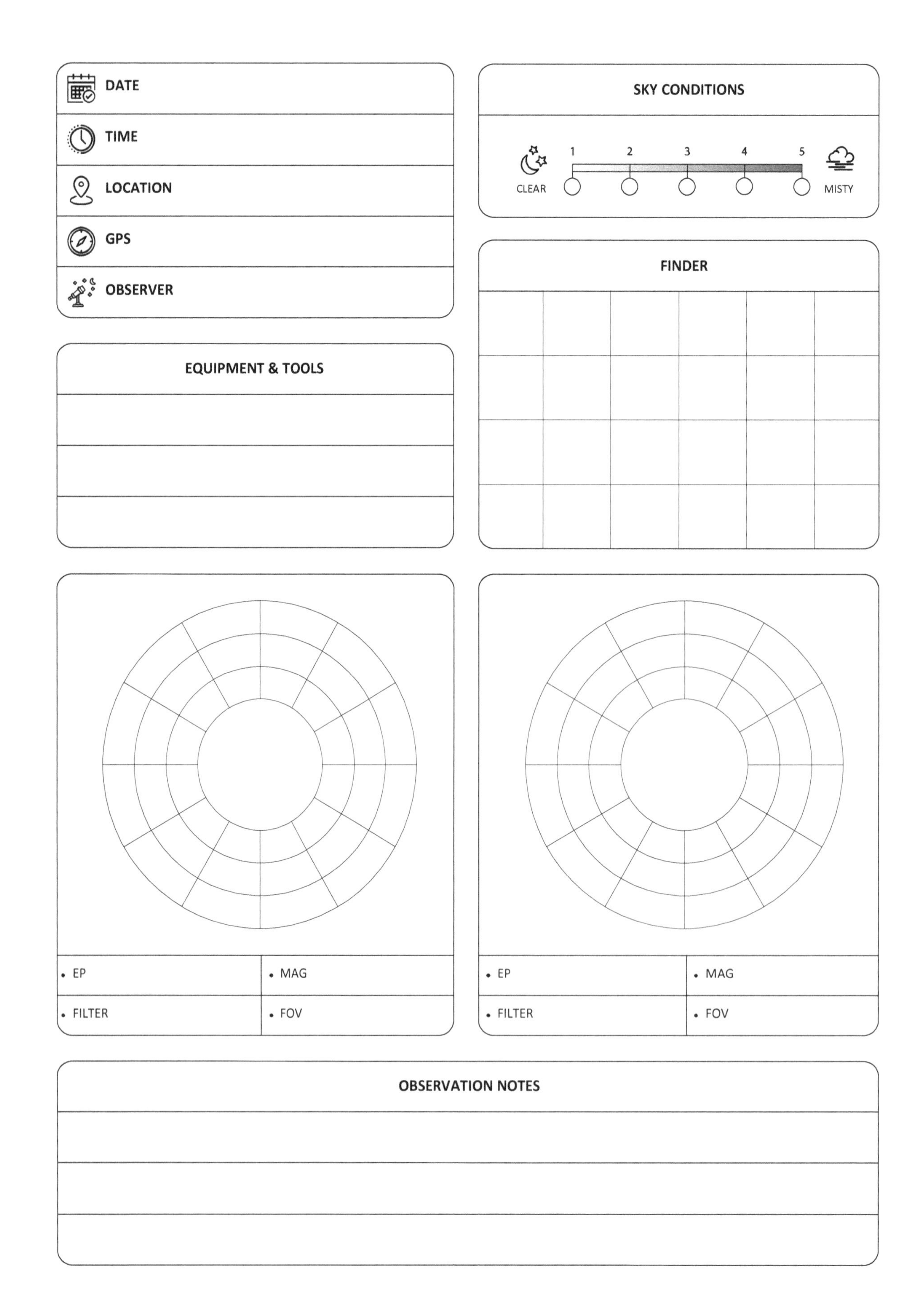

DATE
TIME
LOCATION
GPS
OBSERVER
SKY CONDITIONS
1
2
3
4
5
CLEAR
MISTY
FINDER
EQUIPMENT & TOOLS
EP
MAG
FILTER
FOV
EP
MAG
FILTER
FOV
OBSERVATION NOTES

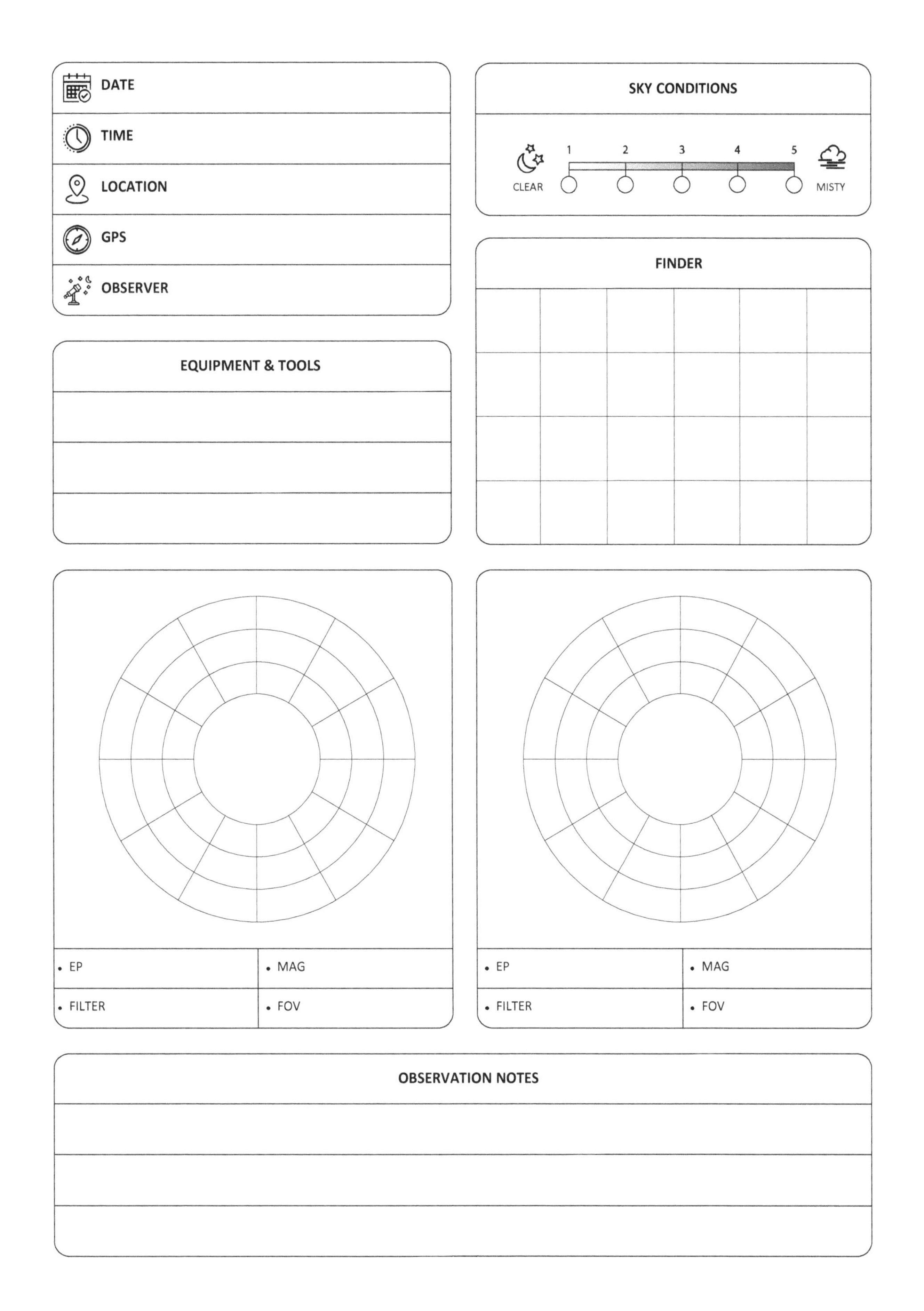
DATE
TIME
LOCATION
GPS
OBSERVER
SKY CONDITIONS
1
2
3
4
5
CLEAR
MISTY
EQUIPMENT & TOOLS
FINDER
EP
MAG
FILTER
FOV
EP
MAG
FILTER
FOV
OBSERVATION NOTES

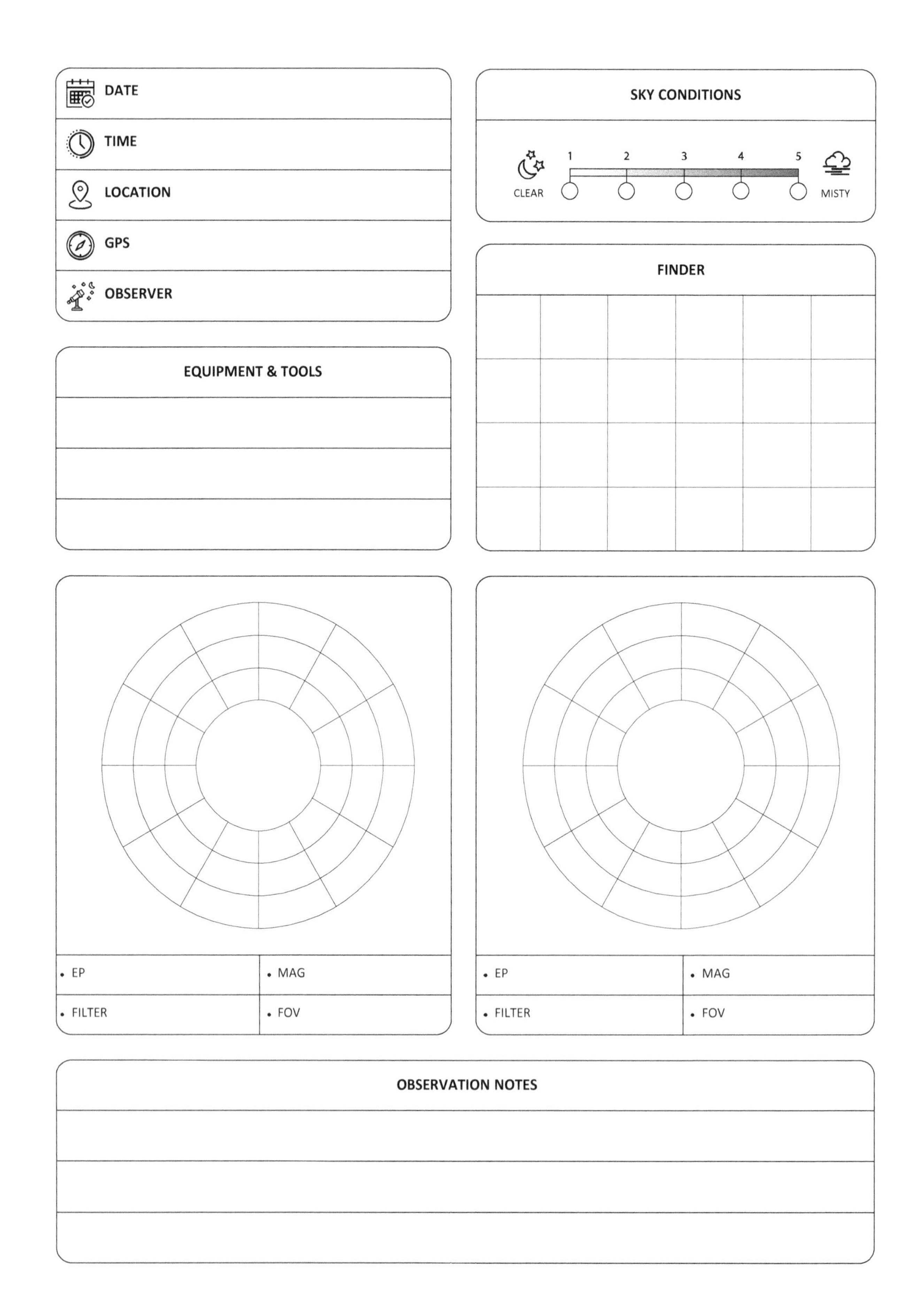
DATE
TIME
LOCATION
GPS
OBSERVER
SKY CONDITIONS
CLEAR
1
2
3
4
5
MISTY
EQUIPMENT & TOOLS
FINDER
EP
MAG
FILTER
FOV
EP
MAG
FILTER
FOV
OBSERVATION NOTES

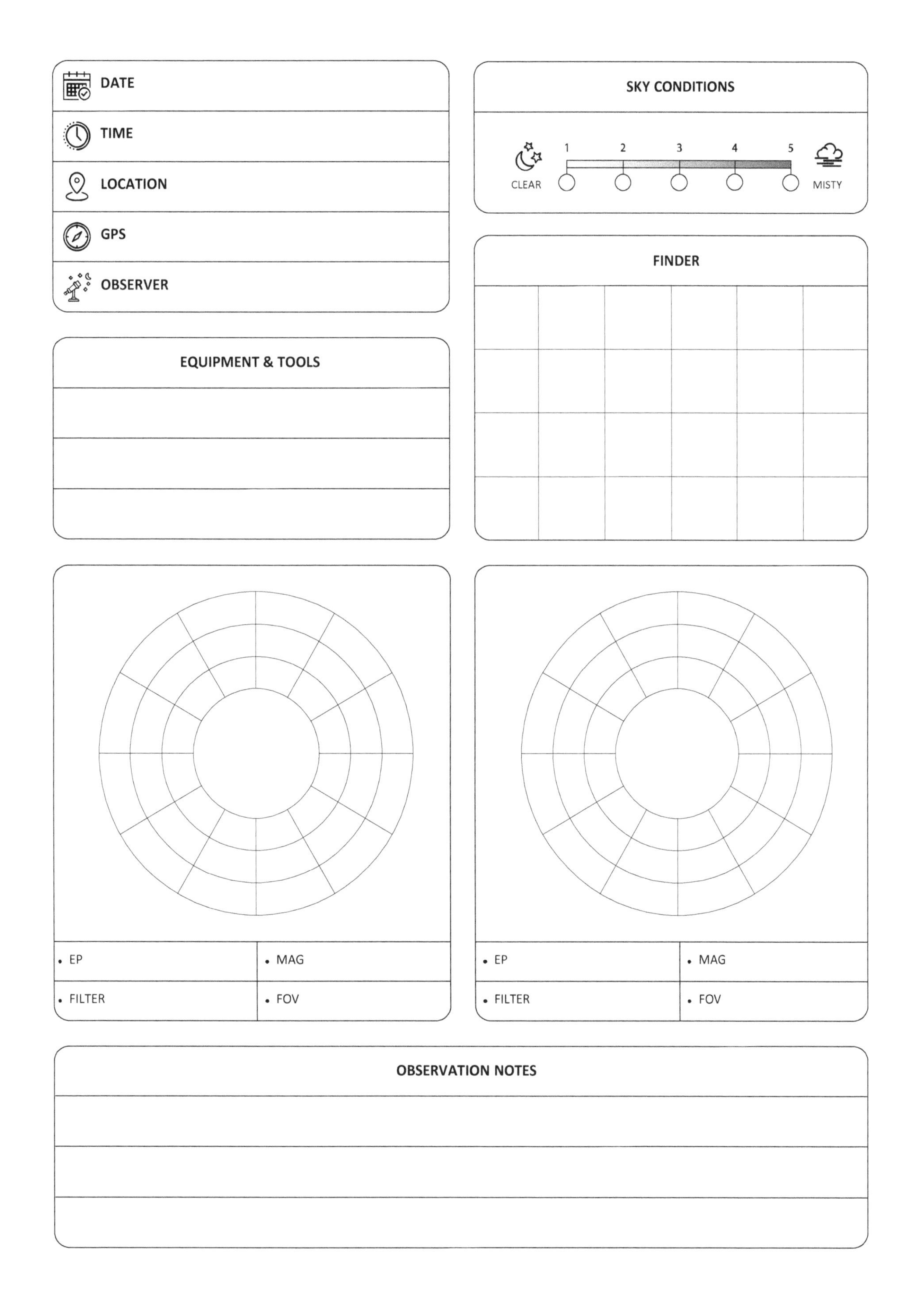

DATE
TIME
LOCATION
GPS
OBSERVER
SKY CONDITIONS
1
2
3
4
5
CLEAR
MISTY
FINDER
EQUIPMENT & TOOLS
EP
MAG
FILTER
FOV
EP
MAG
FILTER
FOV
OBSERVATION NOTES

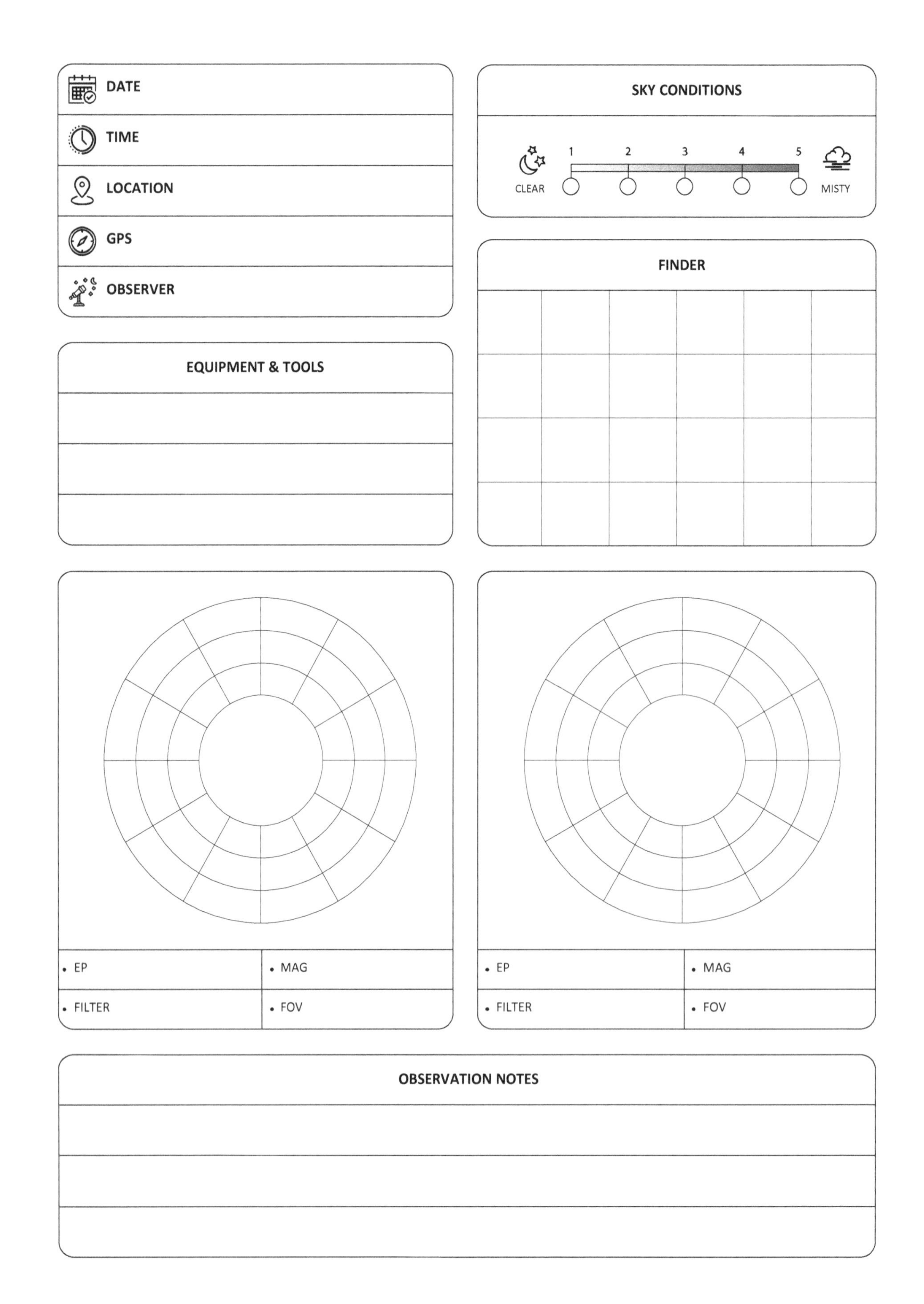
DATE
TIME
LOCATION
GPS
OBSERVER
SKY CONDITIONS
1
2
3
4
5
CLEAR
MISTY
FINDER
EQUIPMENT & TOOLS
EP
MAG
FILTER
FOV
EP
MAG
FILTER
FOV
OBSERVATION NOTES

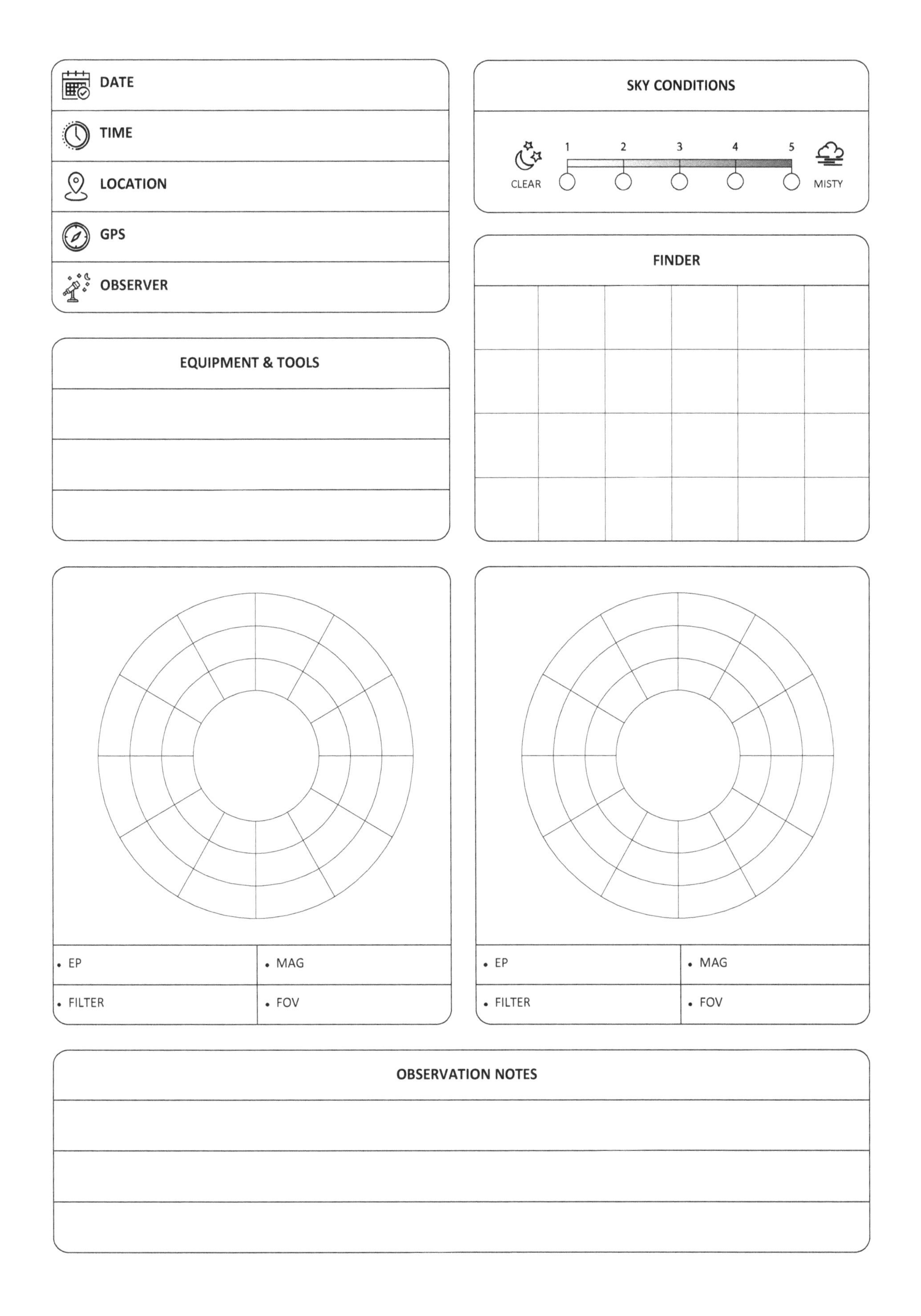

DATE
TIME
LOCATION
GPS
OBSERVER
SKY CONDITIONS
1
2
3
4
5
CLEAR
MISTY
FINDER
EQUIPMENT & TOOLS
EP
MAG
FILTER
FOV
EP
MAG
FILTER
FOV
OBSERVATION NOTES

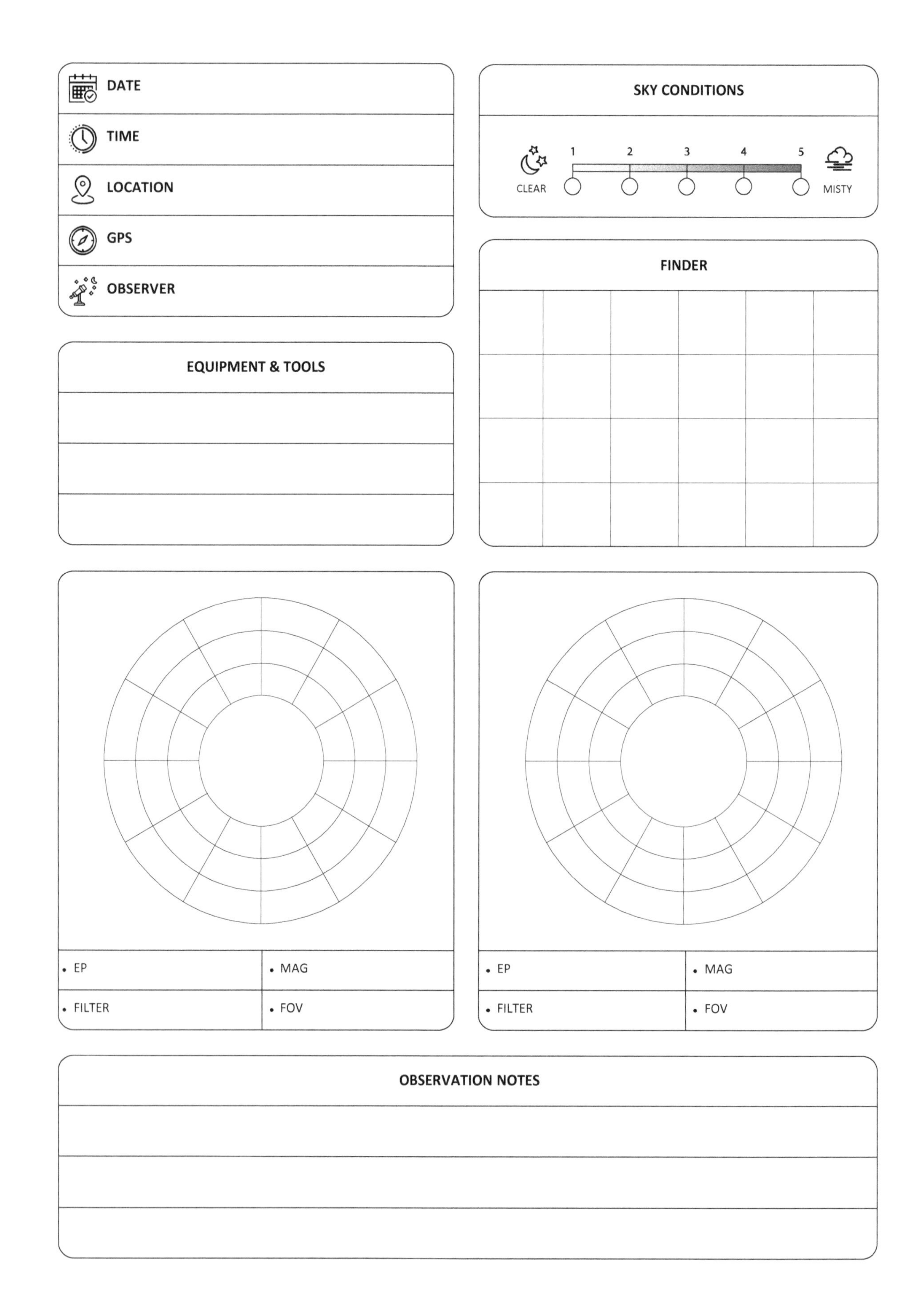
DATE
TIME
LOCATION
GPS
OBSERVER
SKY CONDITIONS
1
2
3
4
5
CLEAR
MISTY
EQUIPMENT & TOOLS
FINDER
EP
MAG
FILTER
FOV
EP
MAG
FILTER
FOV
OBSERVATION NOTES

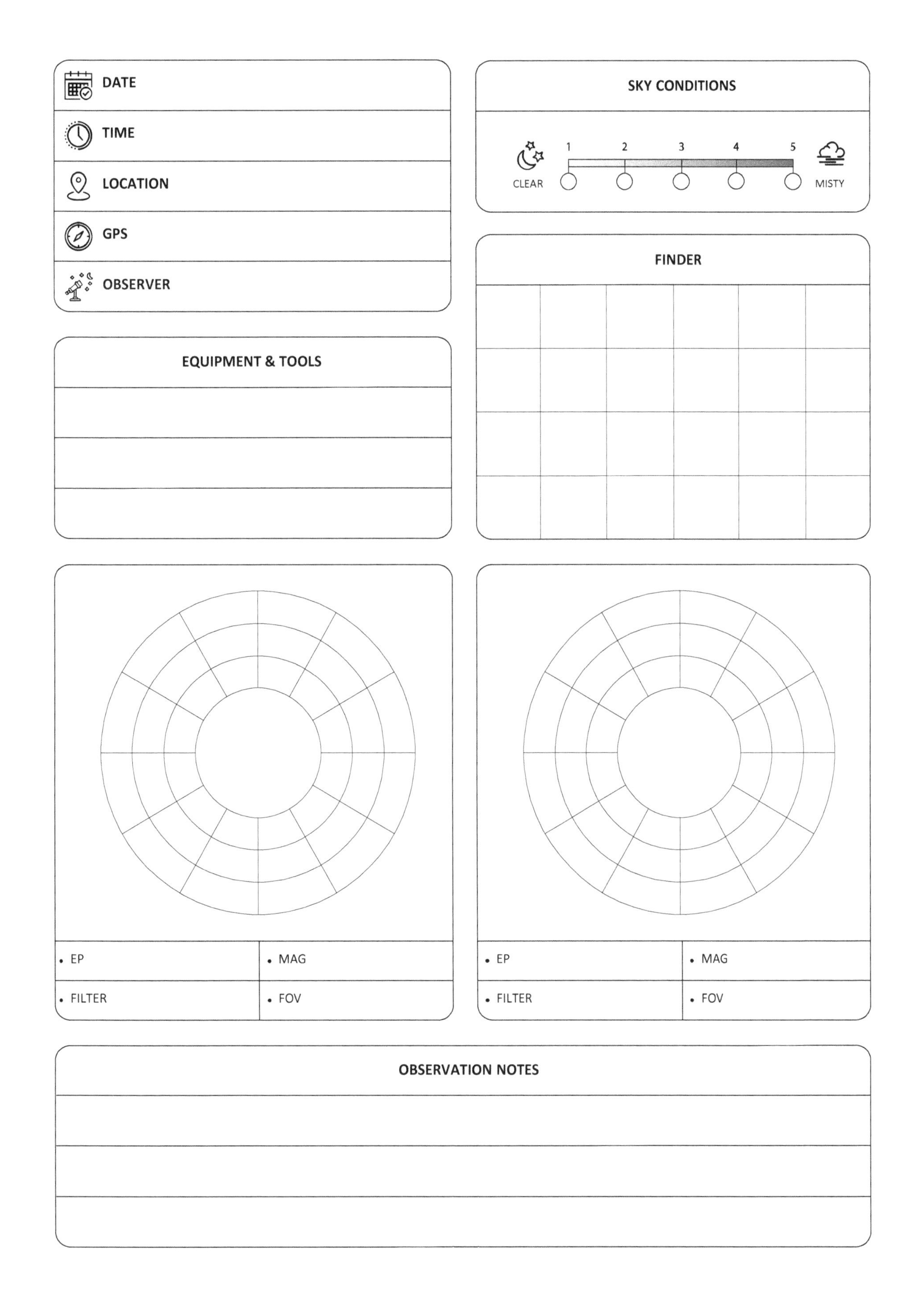
DATE
TIME
LOCATION
GPS
OBSERVER
SKY CONDITIONS
1
2
3
4
5
CLEAR
MISTY
FINDER
EQUIPMENT & TOOLS
• EP
• MAG
• FILTER
• FOV
• EP
• MAG
• FILTER
• FOV
OBSERVATION NOTES

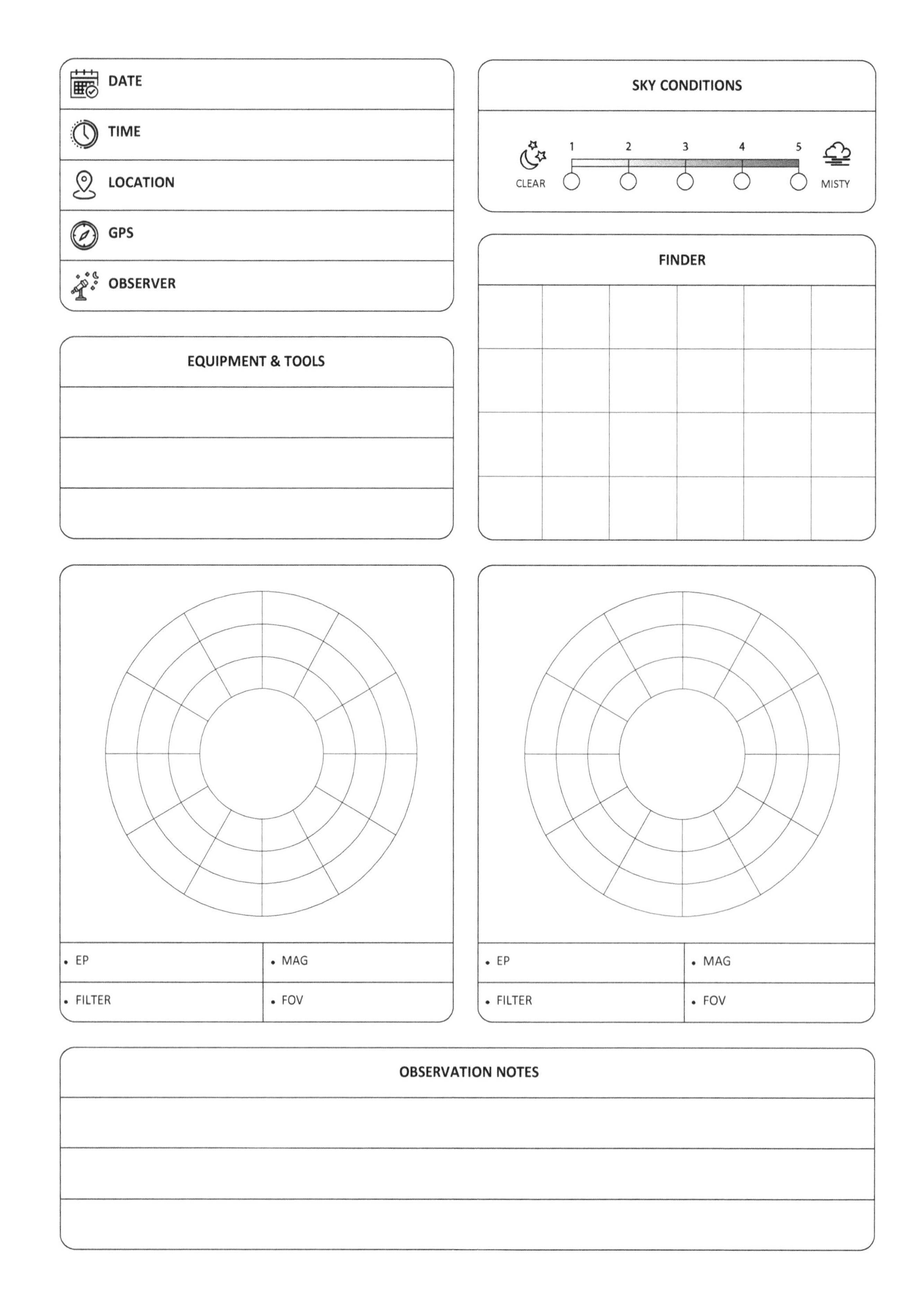
DATE
TIME
LOCATION
GPS
OBSERVER
SKY CONDITIONS
1
2
3
4
5
CLEAR
MISTY
FINDER
EQUIPMENT & TOOLS
EP
MAG
FILTER
FOV
EP
MAG
FILTER
FOV
OBSERVATION NOTES

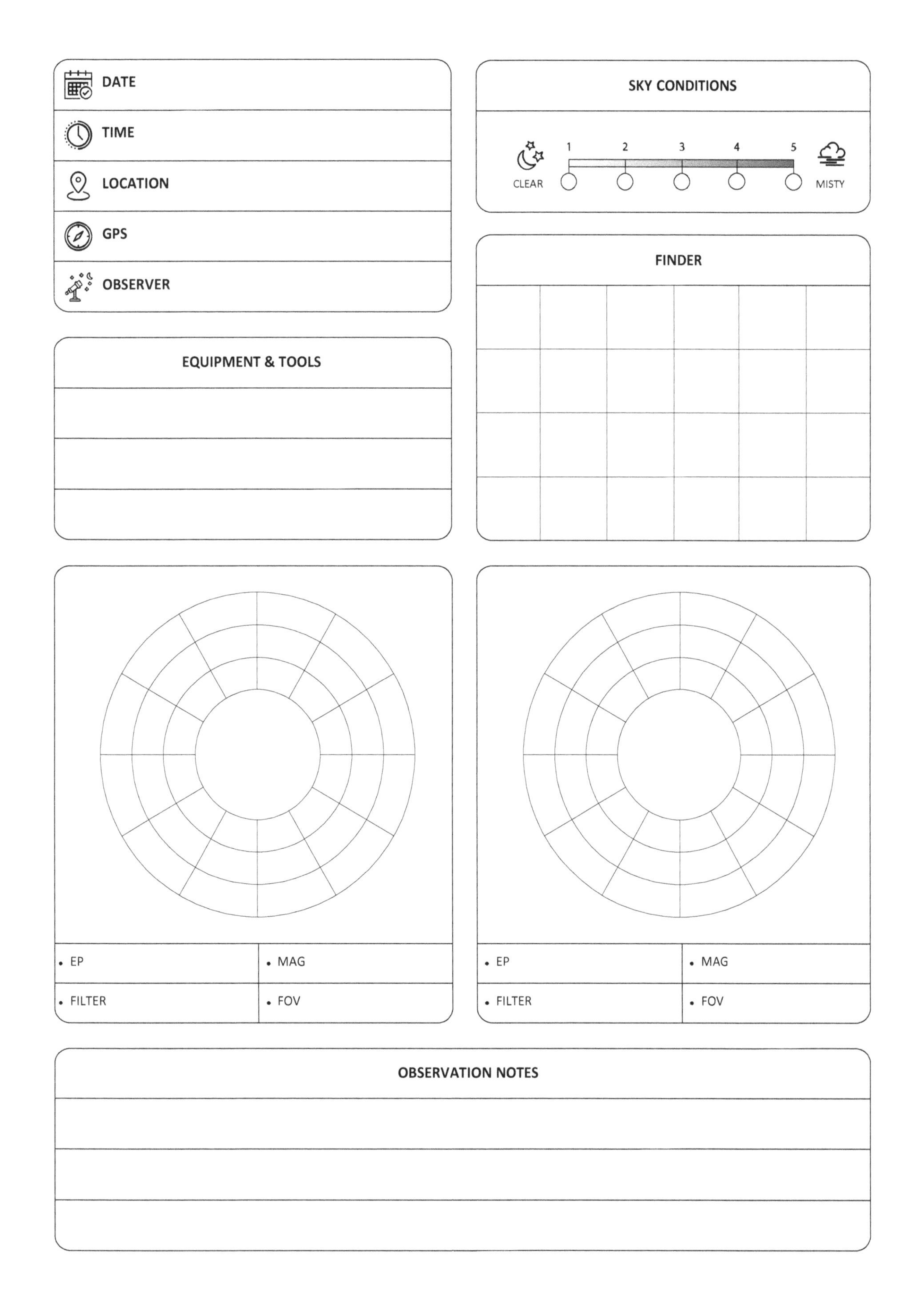

DATE
TIME
LOCATION
GPS
OBSERVER
SKY CONDITIONS
1
2
3
4
5
CLEAR
MISTY
FINDER
EQUIPMENT & TOOLS
EP
MAG
FILTER
FOV
EP
MAG
FILTER
FOV
OBSERVATION NOTES

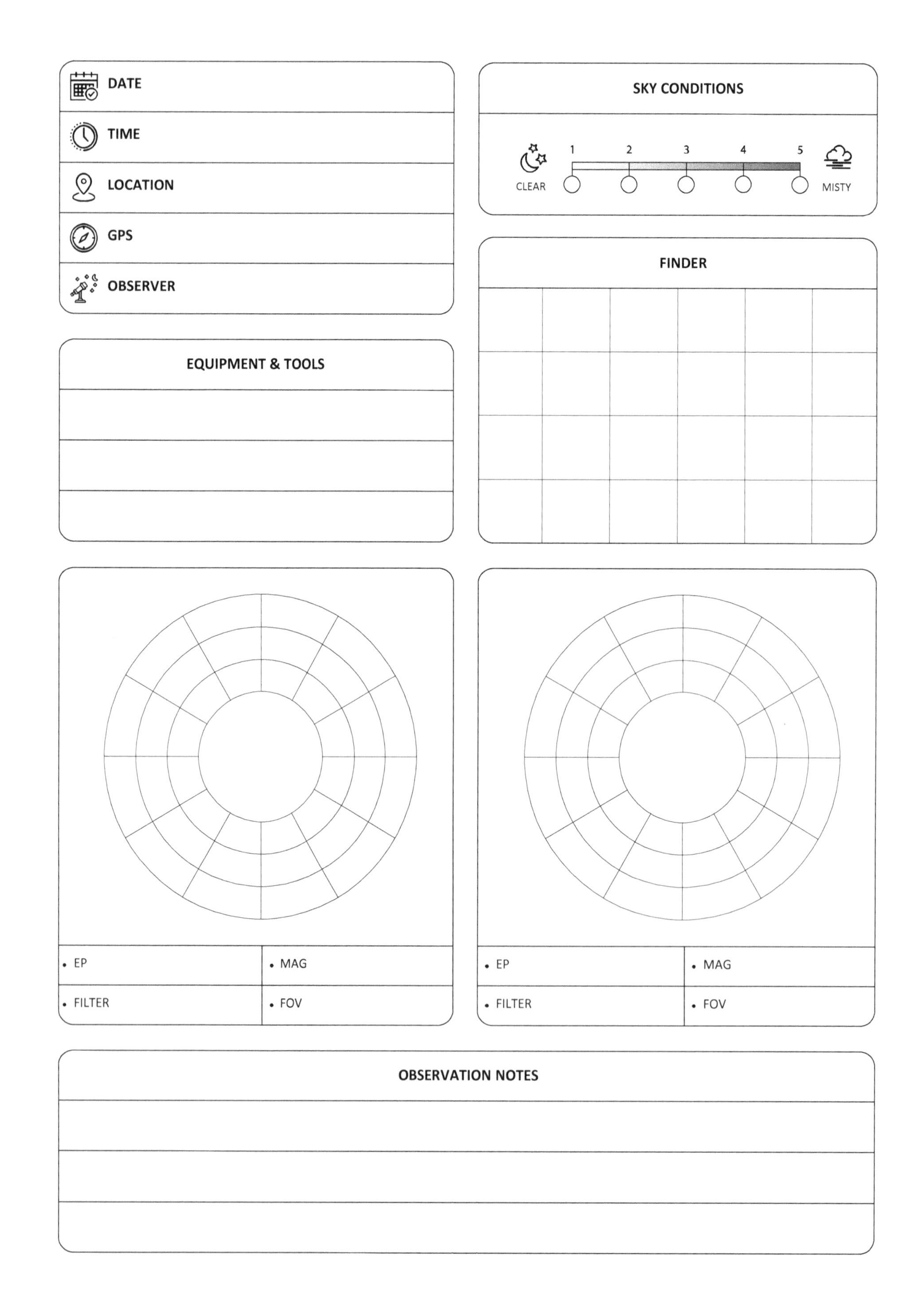

DATE
TIME
LOCATION
GPS
OBSERVER
SKY CONDITIONS
1
2
3
4
5
CLEAR
MISTY
EQUIPMENT & TOOLS
FINDER
EP
MAG
FILTER
FOV
EP
MAG
FILTER
FOV
OBSERVATION NOTES

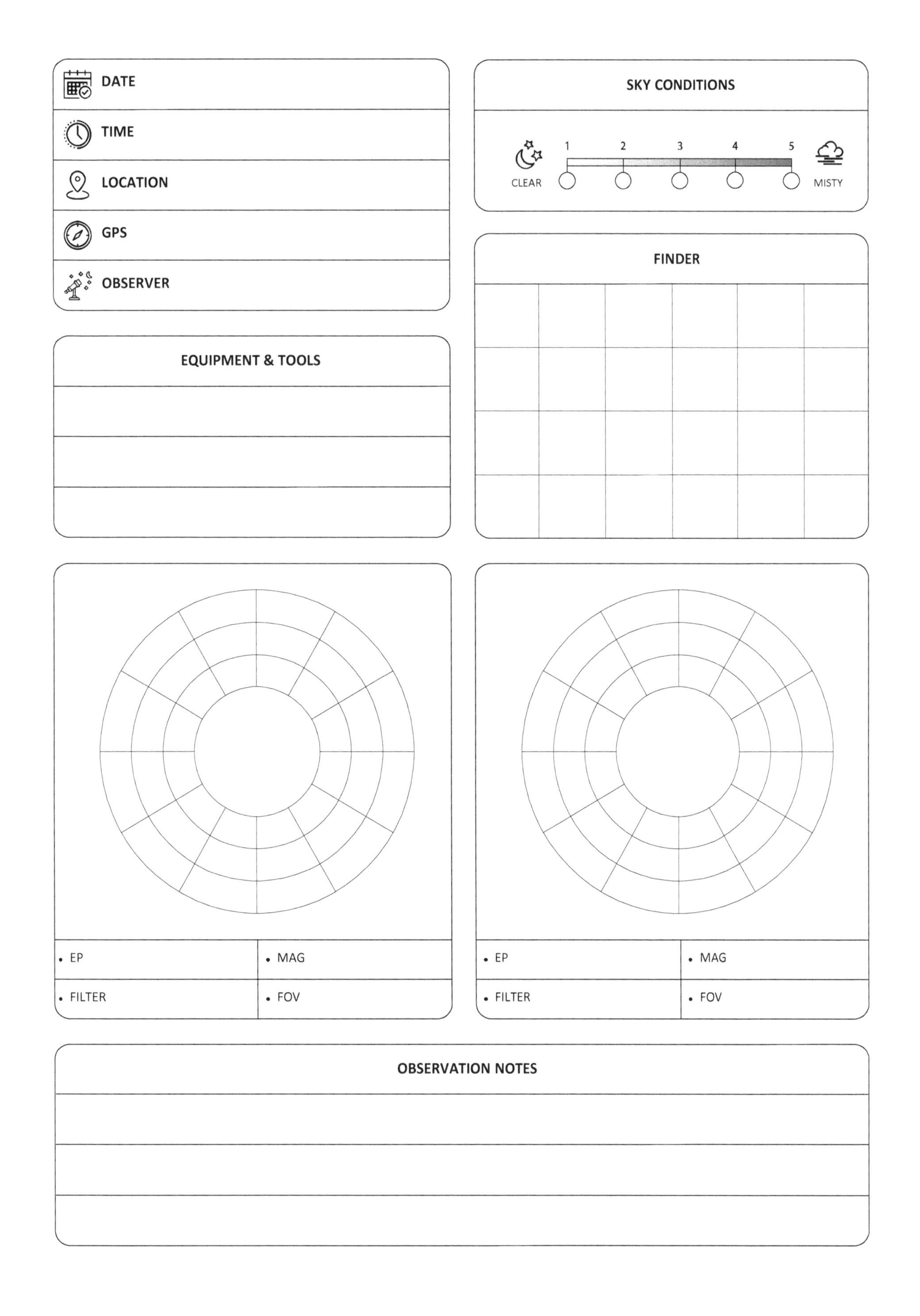
DATE
TIME
LOCATION
GPS
OBSERVER
SKY CONDITIONS
1
2
3
4
5
CLEAR
MISTY
FINDER
EQUIPMENT & TOOLS
EP
MAG
FILTER
FOV
EP
MAG
FILTER
FOV
OBSERVATION NOTES

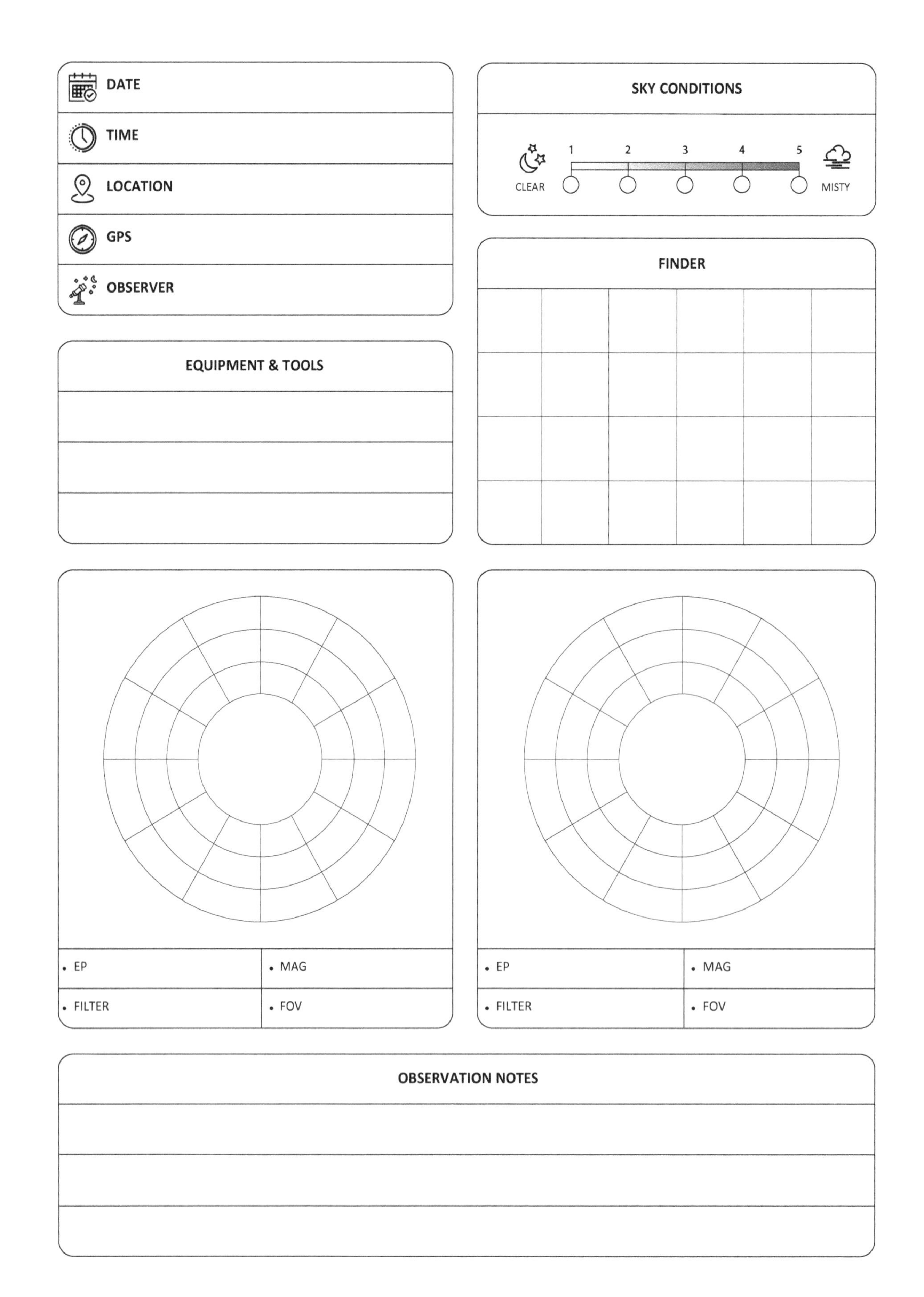
DATE
TIME
LOCATION
GPS
OBSERVER
EQUIPMENT & TOOLS
SKY CONDITIONS
1
2
3
4
5
CLEAR
MISTY
FINDER
EP
MAG
FILTER
FOV
EP
MAG
FILTER
FOV
OBSERVATION NOTES

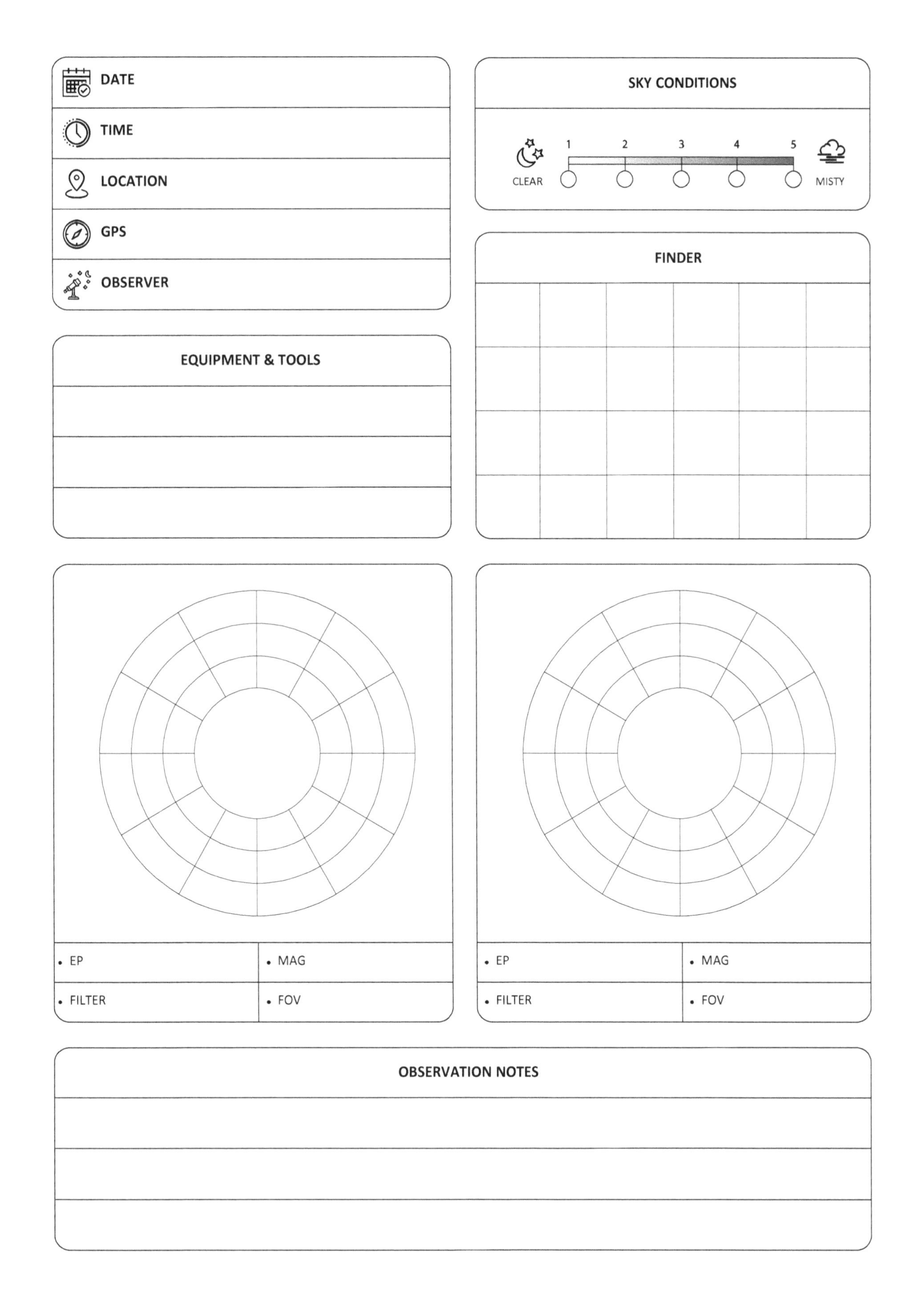
DATE
TIME
LOCATION
GPS
OBSERVER
SKY CONDITIONS
1
2
3
4
5
CLEAR
MISTY
EQUIPMENT & TOOLS
FINDER
• EP
• MAG
• FILTER
• FOV
• EP
• MAG
• FILTER
• FOV
OBSERVATION NOTES

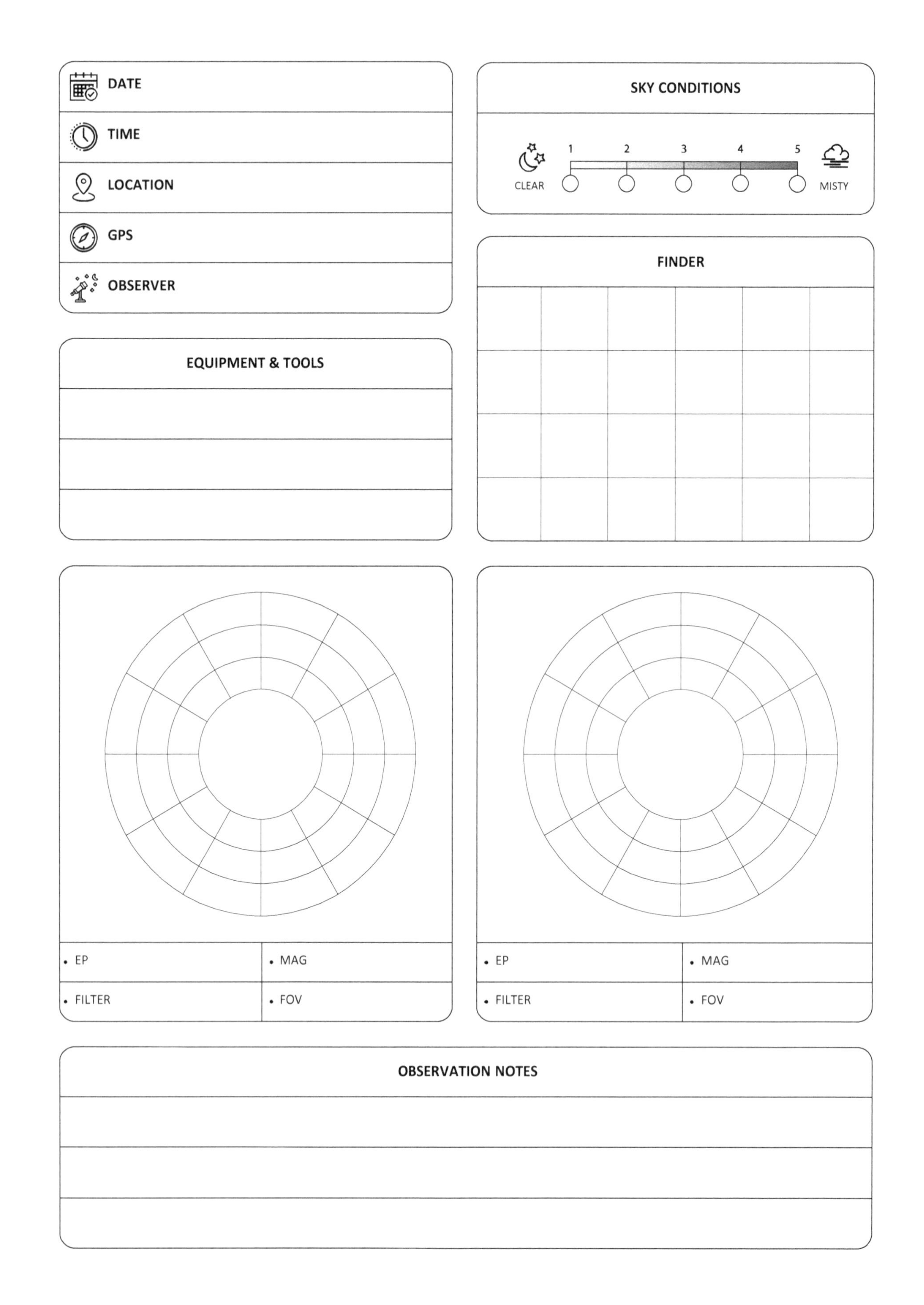
DATE
TIME
LOCATION
GPS
OBSERVER
SKY CONDITIONS
1
2
3
4
5
CLEAR
MISTY
FINDER
EQUIPMENT & TOOLS
EP
MAG
FILTER
FOV
EP
MAG
FILTER
FOV
OBSERVATION NOTES

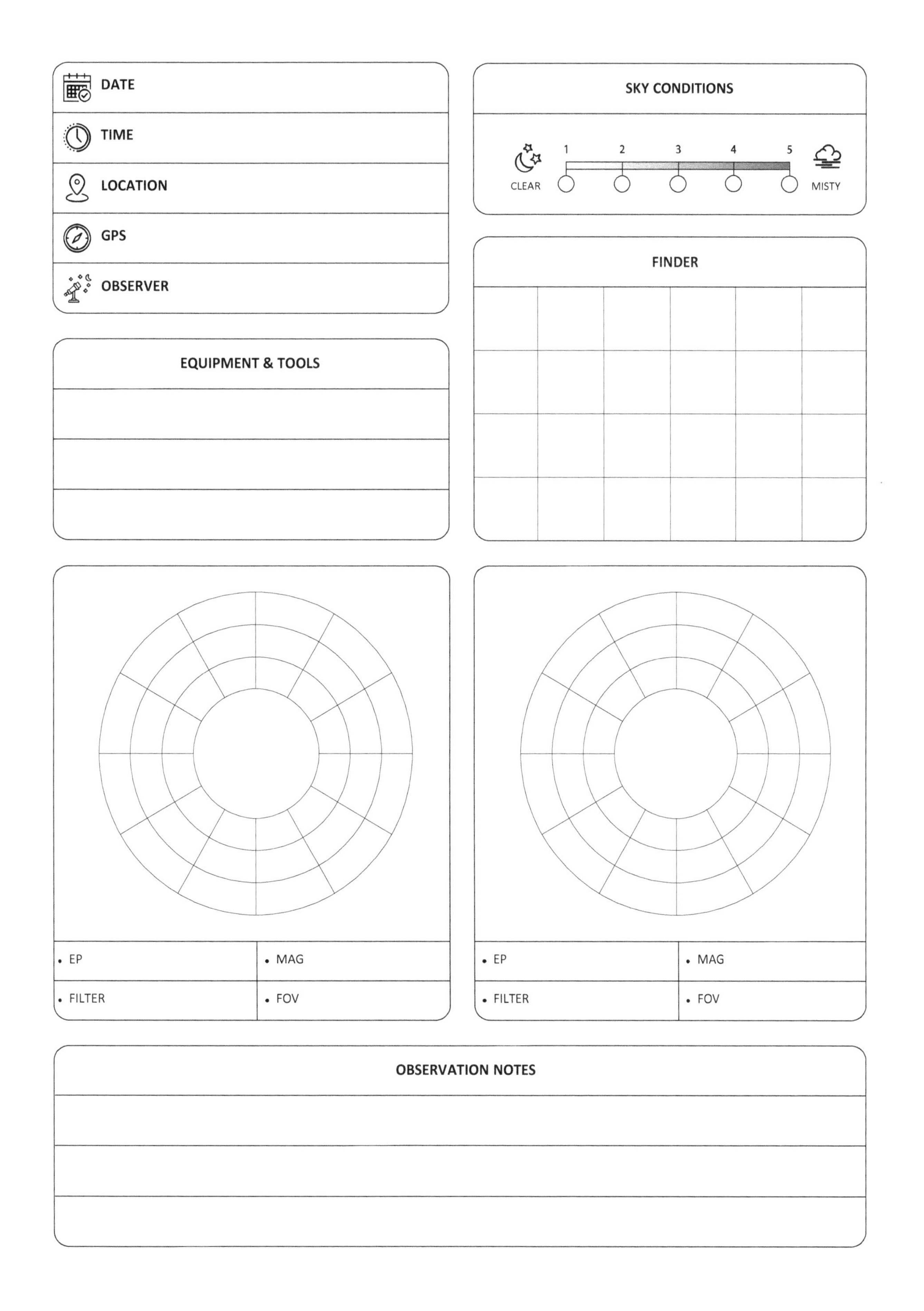
DATE
TIME
LOCATION
GPS
OBSERVER
SKY CONDITIONS
1
2
3
4
5
CLEAR
MISTY
FINDER
EQUIPMENT & TOOLS
EP
MAG
FILTER
FOV
EP
MAG
FILTER
FOV
OBSERVATION NOTES

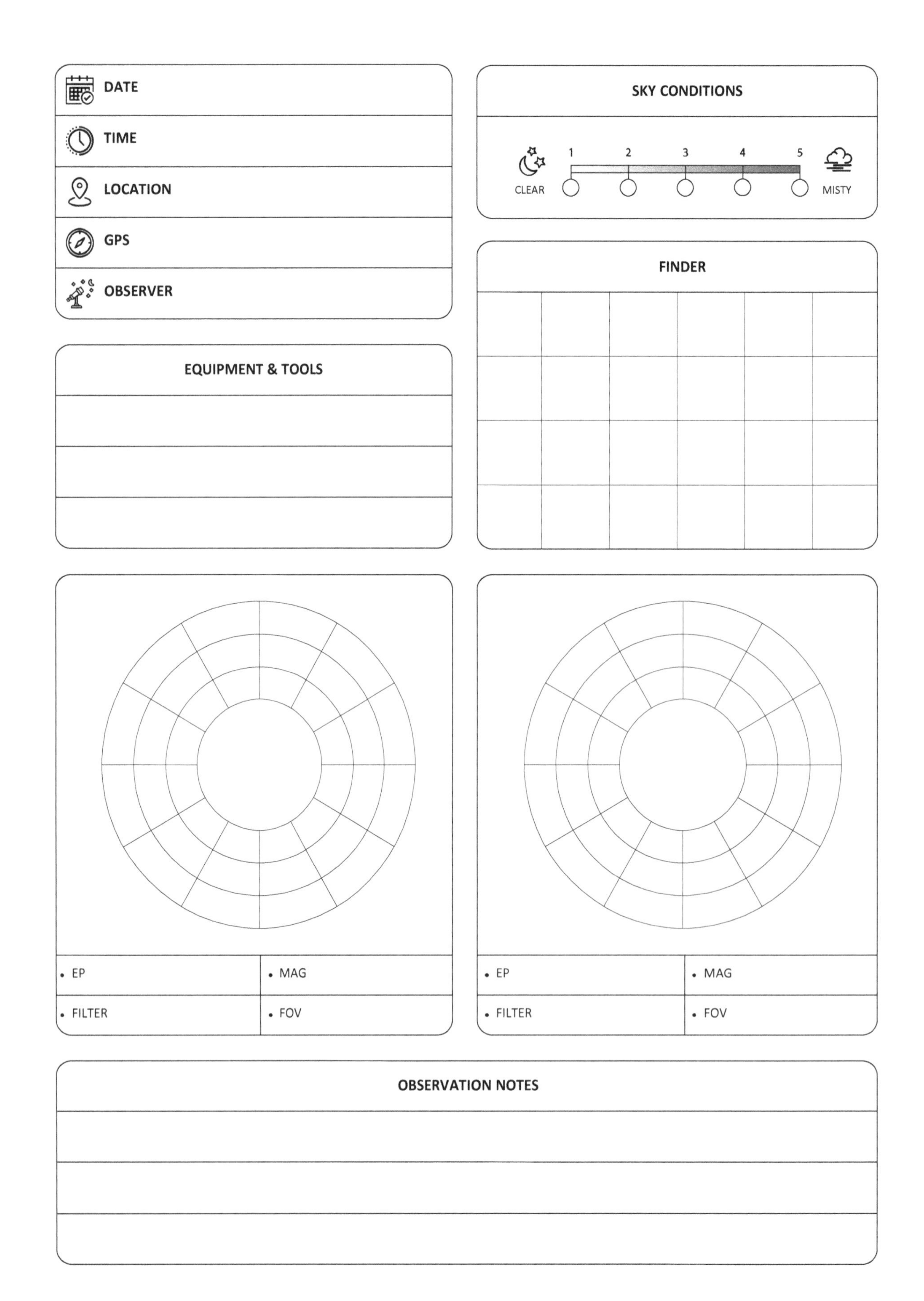
DATE
TIME
LOCATION
GPS
OBSERVER
SKY CONDITIONS
1
2
3
4
5
CLEAR
MISTY
FINDER
EQUIPMENT & TOOLS
EP
MAG
FILTER
FOV
EP
MAG
FILTER
FOV
OBSERVATION NOTES

Lightning Source UK Ltd.
Milton Keynes UK
UKHW032143090223
416681UK00014B/3271